S0-ATD-838

ESSAYS ON THE PHILOSOPHY OF FRED SOMMERS

In Logical Terms

George Englebretsen

Problems in Contemporary Philosophy
Volume 21

The Edwin Mellen Press
Lewiston/Queenston/Lampeter

Library of Congress Cataloging in Publication Data

This volume has been registered with The Library of Congress.

This is volume 21 in the continuing series
Problems in Contemporary Philosophy
Volume 21 ISBN 0-88946-322-0
PCP Series ISBN 0-88946-325-5

A CIP catalog record for this book
is available from the British Library.

For information contact

The Edwin Mellen Press
Box 450
Lewiston, N.Y.
USA 14092

The Edwin Mellen Press
Box 67
Queenston, Ontario
CANADA L0S 1L0

Edwin Mellen Press, Ltd
Lampeter, Dyfed, Wales,
UNITED KINGDOM SA48 7DY

Printed in the United States of America

FOR THE MEMORY OF
MY PARENTS AND SISTER

CONTENTS

FOREWORD

This book makes an unusual and welcome contribution to the field of logic. It is a 'radical anti-consensus piece' if such a term can be applied to so staid a field as logic. Where logical reflection today is in the main formalistic, instrumentalist, and piecemeal, this book is a sustained reflection on the logic of *natural* language, from a realist and systematic point of view. The author manages to combine his careful exposition of this logic with an overview of its history which is an exciting tale in its own right. Let me explain.

In two important senses, Englebretsen is not the inventor of the logic of which he writes, though he no doubt deserves the title of the most dedicated and meticulous expositor of it today. In the first place the logic in question is none other than the so-called 'term logic' usually said to have been invented by Aristotle, taught throughout the middle ages, toyed with by Leibniz, forgotten in the enlightenment and surpassed at last by the great developments in mathematical logic associated with names like Boole, Frege, Russell, Quine. So at least runs the textbook history that the average student of logic would learn today. Term logic figures in the contemporary mind as one of the discarded fashions of science, much like the Ptolemaic system in astronomy. Englebretsen does not claim to invent but only to rehabilitate this logic. And such an effort obviously requires a reassessment of its history, of which the present work provides an outline.

But the logic is not Englebretsen's own in a second way. The book is a sustained and systematic exposition of the life work of Prof. F. Sommers of Brandeis University, whose efforts have revealed the continuity of term logic from Aristotle to Leibniz and also its character as an uncompleted project, with unlimited promise in its application to the logic of natural language. Sommers' work comes at a crucial moment, just as the problems in applying formal mathematical structures to ordinary language are coming to be recognized. Sommers' very unconventional approach, however, has seemed to many to be moving quickly in the wrong direction, toward the 'errors' of the past and he has thus acquired a reputation as the Ishmael of modern logic.

Professor Englebretsen's work is a systematic exposition and defense of Sommers' far-reaching contributions to logic, placing them in the context of a rectified history of the subject. Term logic is a project abandoned prematurely by logicians deceived by the appearance of security which the prestige of mathematics conferred upon mathematical logic. Recent logicians concluded too quickly that term logic was unformalizable, inadequate to reflect many of the actual inference structures of ordinary language, etc. The work of Sommers has demonstrated these claims to be false in the most appropriate way possible, by constructing a term logic of which they do not hold. Moreover Englebretsen has shown that Sommers' reply on behalf of term logic is not a mere riposte; it is a "programme" of logic in the fullest sense. It contains a rigorously presented theory not just of the syntax, semantics and rules of inference for a term logic, but also a modal logic, a theory of predication, identity, singular terms, categories and ontology. In the reading of this book it is impossible not to get the idea that here is a vital programme for logic which is deserving of careful consideration and which is bound to lead to a re-evaluation of the traditional dogmas of mathematical logic.

Graeme Hunter
Department of Philosophy
Ottawa University

INTRODUCTION

None will deny that throughout its long history logic has had its ups and downs. Great advances and insights are accumulated over relatively short periods of time only to be followed by gradual decline and even hostility. Why should this be so for logic, in contrast to other sciences such as physics or biology? The answer, I believe, lies in the fact that logic has always been faced with a dilemma which it has not been able to resolve. Thus, it tends to oscillate over time between two apparently incompatible views of itself. On the one hand, logic is seen as an attempt to account for the kind of natural, everyday reasoning in which we all engage. Consequently, the rules of inference which it expects to formulate must be self-evident to a substantial degree, and they must apply to propositions expressed in the medium of natural language. On the other hand, as a science, logic is seen as an attempt to give the fullest and most rigorous account possible of inference in general. Unfortunately, the results of following the first impulse are often systems of logic with quite limited scope--relevant but powerless. But the second impulse tends to result in systems which are great in power and scope but far too complicated and esoteric to be of ready application to any problems at hand--powerful but irrelevant.

Now it seems to me upon looking at the history of logic that the ascent up one of its mountains is always instigated by an attempt at a relevant logic. Aristotle began with the goal of contriving a logic adequate to the needs of teaching theoretical sciences such as metaphysics and physics. Leibniz's initial goal was a logic adequate to the demands of learning in general. For Frege the correct system of logic would account for the kinds of inferences which characterize proofs in mathematics. The peaks of these mountains (some much broader than others), however, are usually seen to be characterized by systems of great complexity, whose relevance to the task for which they were first invisaged has either been abandoned or radically distorted. The plunge down the other side and the long valley which follows are always the result of a reaction on the part of philosophers (and some scientists) to the logician's failure of relevance. Leibniz's upward efforts are almost lost in the broad deep valley of the Seventeenth and Eighteenth Centuries. That low point for logic was the result of early modern philosophers' attempts, prompted by the new science, to abandon the complex, dense, archane logic of the late scholastics.

The past century has witnessed logic's most rapid and highest ascent. The modern mathematical logician, standing, ultimately, on Frege's shoulders, commands an unprecedented and unobstructed view. Yet, already, in spite of the great power, scope and beauty of the new logic, there are those who have begun to demand its applicability outside mathematics and to attack its complexities. The modern mathematical logician, of course, seldom bothers to look down (and never back to the valleys and lower hills of the past), but keeps both eyes on still higher grounds.

Well, I suppose I've carried this metaphor far enough. The point is that we are now once more at a time in the history of logic when the forces devoted to building a system of logic which is, above all, powerful are beginning to lose at least a bit of ground to those who demand relevance (in the sense at least of applicability to the need for understanding ordinary reasoning as exhibited in the use of natural language inferences). The calls to abandon the high ground are heard louder each day from the "informal" logicians, the "relevance" logicians, the teachers of "critical thinking." And, face it, most philosophers, certainly all nonanalytic philosophers, have already abandoned mathematical logic to the mathematicians.

But need logic's next epoch be one of little rigor, oversimplification, too-easy accessibility and weakness? Unless we are fatalists, we are free to follow Santayana's advice to avoid repeating the past. I believe it is possible to glean a large number of lessons from logic's history. I also believe that it is possible to overcome the dilemma which has for so long driven logic on its uneven course. It is quite possible to formulate a system of logic which is at once both powerful and rigorous, fit for the demands of the strictest formalist, and at the same time simple and natural, applicable to the philosopher's need to account for reasoning in ordinary life. In fact I can go farther. The task is already well underway. An understanding of the ancient lessons of logic, an ability to construct formal systems of rigor and power, and an appreciation and respect for natural language as a proper medium in which to conduct the business of everyday reasoning are all combined in the work of Fred Sommers.

I have hoped for many years that a book introducing, explicating and summarizing Sommers' wide-ranging and original ideas would be produced by someone of greater abilities than myself. Unfortunately none has been forthcoming. My sincere belief is that Sommers' work is deserving of a much larger audience than it has so far enjoyed. It is for this reason that I have decided to offer here these essays, with the hope that some sparks of interest may fly even from a less-than-perfect flint.

Sommers' published work can be divided into three distinct but related parts. The first part consists of his earliest essays on the structure of ordinary language. He built and defended a theory of structural isomorphism between language and ontology (called the "tree theory") beginning with his 1959 essay "The Ordinary Language Tree." Most of his papers until the mid-1960's were devoted to this project. During that period he published two brief essays concerned with logic proper ("Truth-functional Counterfactuals" 1964, and "Truth-value Gaps: A Reply to Mr. Odegard" 1965). But beginning in the late 1960's he produced a long series of essays and his book, *The Logic of Natural Language* (1982) in which he constructed and explored a logic of terms and an algorithm for manipulating it (the "new syllogistic"). Finally, Sommers has recently formulated a fascinating and original theory of truth (the "constitutive correspondence theory"), though his initial discussion of truth actually came in his 1969 "On Concepts of Truth in Natural Languages."

These three theories: the tree theory, the new syllogistic and the constitutive correspondence theory, are all theories which deal with language in one way or another. The three are closely related. Indeed, they can be thought of as three parts of a single overall theory of language. But there is one very strong thesis which can be found at the heart of these three subtheories, and it is probably this rather than any other shared feature which accounts for their mutual cohesion. The thesis I refer to is this: contrary to the generally accepted view among logicians today (but in accord with many linguists), the negation of a term in natural language cannot be reduced to the negation of a sentence. In effect, the thesis is that there is a fundamental, logical distinction between term negation and sentence negation (denial). Sommers often refers to this thesis as the "denial/negation distinction." It is important to realize that Sommers is not, nor does he claim to be, the author of this thesis. It is implicit throughout the entire corpus of traditional, pre-Fregean logic. What Sommers has done is this: he has, on the basis of that old thesis, constructed original and suggestive theses concerning a wide variety of aspects of language.

One of the reasons why the work of Fred Sommers has attracted my attention and admiration for so many years is that it constitutes a philosophical program. More importantly, it constitutes an *explicit* philosophical program. The idea that a philosopher should set out to construct a program for doing philosophy, attacking philosophical issues, is commonplace among nonanalytic philosophers. But twentieth-century analysts seem afraid of, or even hostile to, the idea. Admittedly, a philosophical program can be extrapolated from the work of those philosophers who have approached a sufficiently large variety of philosophical problems from a sufficiently small number (ideally one) of starting points. All the very good analysts, if they produce enough work, offer such *implicit* philosophical programs. Sometimes they even become aware of at least some facets of the program and then overtly exploit these in further work. Often a philosophical program is nothing more than a commitment to a particular principle which is seen as relevant to a large number of philosophical problems. Thus Ryle's notion that category mistakes lie at the heart of most, if not all, philosophical puzzles might have constituted his program. At one time Russell's program was embodied in his thesis that philosophy is logic. An obvious famous example of a nonphilosophical program today (though it has numerous philosophical applications) is Chomsky's program for grammar. Notice that one can think of all the changes, especially in the last few years, that his work has undergone as refinements in the original program rather than as new programs. Like Chomsky's program for grammar, Sommers' philosophical program has been explicit and dynamic. I could add that, again with Chomsky, Sommers' program is rich. Let me explain each of these features.

Moore once said that in doing philosophy he simply waited for some philosopher to claim something and then he (Moore) set to work analyzing the claim. This *ad hoc*, retail approach had much to recommend itself to Moore over the wholesale approach he had encountered as a student. I once met a philosopher in Ontario who carried Moore's attitude to a ridiculous extreme. He informed me that once a month or so he would go to the library, read the first article in the most recent issue of the *Journal of Philosophy*, and then spend the rest of the month trying to solve the problem dealt with therein! From the beginning of his philosophical career Sommers set out not to solve a philosophical problem but to initiate a program which could then be used to solve a large number of such problems. This is what I mean by saying that his program has been explicit. It is not something gleaned from a large repertoire of solved problems. Instead, he deliberately attempted to construct a program which could then be applied to the problems.

Sommers' philosophical program is dynamic in the sense that he has continued to refine it over the years. It has been refined in two ways. First, he has tended to treat it as follows. He formulates a thesis, applying it then to various problems. In light of how well the thesis deals with such problems he modifies it in order to improve its performance. Secondly, the presentation of problems just beyond the scope of the thesis has elicited extensions of it--the program has grown.

And it is because it has grown so that Sommers' program is so rich. It is rich in the sense that it is applicable to a very wide range of related but distinct kinds of philosophical problems. His philosophical program is "big."

In spite of the richness and dynamism of Sommers' program it has one feature which makes it practical and teachable. Underlying what may seem to be its complexity and looseness is a simplicity in rigor which is nothing less than beautiful.

All of these essays deal directly (or nearly so) with ideas first instilled in me by reading or talking with Sommers. And while he is the inspiration for them, he is responsible for none of my shortcomings in attempting to reach the high standard of clarity, rigor and originality which he has set in his own work. In addition to my deep and long-standing gratitude to Fred Sommers, I gratefully acknowledge the suggestions, help, constructive criticism, patience and encouragement of Bill Shearson and Harvey White. I also want to express my appreciation to several former students who have listened to my ideas about logic and language over the past twenty years and in spite of this have continued to be lovers of wisdom: J. Fjeld, G. Hunter, J. Blevins, T. Coats, S. Milner, P, Smith, R. Besner and O. Mercure. Finally, I want to take this opportunity to thank my former teachers, Charles Sayward and Steve Voss, who introduced me to Sommers, his work, and much else besides.

Chapter I

A REINTRODUCTION TO SOMMERS' TREE THEORY*

Fred Sommers' work on ontology (the study of categories) has earned far less critical attention than it deserves. Sommers worked out his theory in a series of papers and addresses during the sixties;[1] and while some attention was paid to this work at the time, little or none has been paid since.[2] Just recently a special issue of the *Monist* was devoted to the topic of categories.[3] Notwithstanding the importance and high quality of the essays there, it is remarkable that Sommers' highly original and suggestive ideas on categories are never mentioned by any author. One reason of course for the decline of interest in Sommers' ontology is his own shift of interest, since the late sixties, to the logical work for which he has received much more attention.[4]

I believe Sommers' theory of categories is well worth retrieving. This essay is intended as a reintroduction to that theory. This is a valuable exercise for at least three reasons. Ontological questions are of central importance to philosophers, and any original, well-formulated ontological theory ought to be given a fair hearing. Moreover, Sommers' particular theory is replete with insights and suggestions for the management of a wide variety of issues in metaphysics, epistemology, the philosophy of mind, and theology. And, finally,

Sommers' category theory plays a key role in his new logic, a fact substantiated by chapter thirteen of his recent book, *The Logic of Natural Language.*[5] In the preface to that book he says that he had first planned a book on categories but was putting it aside temporarily for this book on logic. He still intends a book-length treatment of categories. If he completes such a project this essay might serve as a kind of advanced introduction. At the very least I hope that it can serve to prod some readers to return to Sommers' own work, which began more than three decades ago.

Classes and Categories

The world is filled with all sorts of things. And language is full of all sorts of expressions. Most of the expressions (terms) can be used to characterize the things. Thus, 'red' characterizes red things, 'married' characterizes married things, and 'in the garden' characterizes things in the garden. The relatively few expressions which are not terms are *formatives*, used to form sentences, sentence parts, or other expressions. Quantifiers, copulae, conjunctions, and the like are formatives.

Every (sense of a) term determines a kind of thing. The kind of things determined by a term is called the *class* with respect to the term. Thus: the class of red things (i.e., the class with respect to 'red'), the class of married things, the class of things in the garden. More than one term might determine a given class. For example, the class of triangles and the class of three-sided plane figures are the same class. As one might expect, there are far more ways of characterizing things than there are kinds of things.

Terms are true of the things they characterize--false of the things they do not characterize. 'Red' is true of Mars, false of the Moon. We sometimes say that the class of things determined by a term (things of which that term is true) is the *extension* of that term. Mars, but not the Moon, is in the extension of (class determined by) 'red.' There are terms which are true of just one thing, though most are true of more (often very many) things. 'Red' is true of many things. A term like

'U.S. president' is true of a few things. 'P. F. Strawson' is true of just one thing.

So far we have a nice, neat, simple picture. The world is full of lots of things. Things are grouped together, classified in a variety of ways according to their characteristics, and our language is more than adequately supplied with terms for us to use in characterizing those things. Given a class, say the class determined by 'married,' we can say of each thing in the world either that 'married' is true of it (in which case it is in the class) or that 'married' is false of it (in which case it is outside the class). The interesting features of classes and their relationships are explored systematically in the theory of classes (or set theory). Two important features of sets or classes will be of interest to us here. One is that everything in the world belongs to the class of things that are X or to the class of things of which 'X' is not true (for any term 'X'). Given 'red,' say, whatever one might mention is either in the class of red things or in the counter-class of red things (i.e., everything else). In other words, any class and its counter-class divide the things in the world into two mutually exclusive and exhaustive groups. The second feature of classes is this. Any pair of classes is such that all the things in one are also in the other, or some things (but not all) in one are also in the other, or nothing in either belongs to the other. Technically speaking, the three cases are said to be inclusion, intersection (overlap), and exclusion. For example, the class of dogs includes the class of collies, the class of dogs and the class of brown things overlap, and the class of dogs and the class of cats are exclusive, have no common members.

To repeat, a nice, neat, simple picture. If it were the whole picture set theory would suffice for much of ontology. Of course, it is not the whole picture. It is, to be sure, the clearest, most familiar, most used part of the picture. There are problems about mere classifying as the best way of going about ontology. The attempt to overcome them leads to a richer and more subtle way of dividing up the things of the world.

Sommers defines 'ontology' as the "science of categories" ("Types and Ontology," p. 351). More specifically, it is the systematic attempt to say what categories are, how they are determined, and how they are related to one another. Ontology is, of course, nothing new. Plato's theory of forms, Aristotle's *Categories*, Russell's theory of types are three quite obvious examples of attempts at a science of categories. But it is Sommers' own theory which comes closest to a complete, workable theory of categories.

So what *is* a category? It is a group of things, a class--but a special kind of class. We saw above that a class consists of all those things of which some given term is true. The notion *true of* is the source of one of the problems with classes. Consider how we negate. Modern logicians take all the kinds of negation which can occur in a natural language sentence as amounting to the negation of the entire sentence. Thus, 'Sam is unmarried,' 'Sam isn't married,' 'It is not true that Sam is married,' and so on are all parsed à la 'It is not the case that Sam is married.' Nonetheless, traditional, pre-Fregean logicians generally recognized two quite distinct kinds of negation.[6] Terms could be negated (e.g., 'unmarried,' 'non-red') and sentences could be negated. What the modern logician does is equate the negation of the predicate term in a sentence with the negation of the entire sentence. He does this, in part, because otherwise it would be possible to have two sentences of the forms 'S is P' and 'S is not P' both being false--breaking the law of excluded middle. The law is easily preserved by equating 'S is not P' with 'not: S is P.' Yet, if there are cases where sentences of the forms 'S is P' and 'S is not P' are both false, then, to preserve excluded middle, one must follow the traditional logician in distinguishing term-negation from sentence-negation. And there are plenty of such cases: 'The lowest prime number is married'/ 'The lowest prime number is unmarried,' 'Kant is even'/'Kant is uneven.' Notice, for example, that to deny that the lowest prime number is married is not to affirm that it is unmarried.

Let us make a distinction between what a term is true of and what it *spans* ("Types and Ontology," p. 329). A term spans whatever either it or its negation could be used to sensibly characterize. And all the things spanned by a given term constitute the *category* with respect to that term. Thus, while the lowest prime is in the counter-class of 'married,' it is not in the category determined by 'married.' While a term and its negation exclusively and exhaustively divide the world, a term and its negation both determine the same category, which need not exhaust the world. This is because whatever can be sensibly characterized by a term can be sensibly characterized by its negation. Spanning is not defined in terms of truth. While a term and its negation are never true of the same things, they both span the same things. For example, whatever can be sensibly characterized as married can equally well be characterized (though perhaps falsely) as unmarried. It would be false, but nonetheless sensible, to characterize Queen Elizabeth as unmarried. On the other hand, neither 'married' nor 'unmarried' can be used to characterize the lowest prime number. To characterize it in either way would be senseless, nonsense, *category mistaken*.

While classes are defined in terms of truth, categories are defined in terms of spanning (which in turn is defined in terms of sense). A category with respect to a term 'X' will include the class of X things and some (usually not all) of the counter-class of X things. The class of married things has in it me, the Queen, my wife, etc. The counter-class has my son, the Pope, the Moon, Mars, etc. (viz. anything of which 'married' is not true). The category with respect to 'married' has me, the Queen, my wife, my son, the Pope, etc., but not the Moon, Mars, etc. For, while 'married' is false of, say, both my son and Mars, it spans only the former and not the latter.

Another important way in which categories differ from classes is this. Pairs of classes, as we saw, are always related in one of three possible ways: inclusion, overlap, exclusion. Much of Sommers' work has established what he calls the "law of category inclusion."

> If C_1 and C_2 are any two categories, then either C_1 and C_2 have no members in common or C_1 is included in C_2 or C_2 is included in C_1 ("Types and Ontology," p. 355).

In other words, categories, unlike simple classes, can include or exclude one another, but cannot overlap. Sommers and others have shown how important the consequences of this law are for a variety of important metaphysical issues. For example, if one defines an *individual* as a thing belonging to categories, all of which obey Sommers' law, then whatever is in the intersect of two overlapping categories will be a nonindividual. Nonindividuals are generally nonsensical entities (like red numbers, married planets, or talking teacups), and generally to be excluded from the ontology. Others are composites of more basic (ontological) entities. Thus, Descartes took souls to belong to the category of things with respect to 'thinks' and bodies to belong to the category of things with respect to 'extended.' Neither category is included in the other, and persons seem to constitute a common subset of both. Nevertheless, Descartes' theory preserves Sommers' prohibition against category overlap by taking persons as nonindividuals. He took persons to be composites of two other individuals--souls and bodies. Whether we accept or reject Cartesian dualism, we can understand it only in terms of categories--not classes.

There is a special kind of category in Sommers' theory which cannot be overlooked. Given a finite number of categories and Sommers' law, there will be a category which includes all other categories and several which include no other categories. A category which includes all other categories must be determined by a term which spans all individuals. Examples of such terms are 'exist' (and 'not-exist'), 'individual,' 'mentioned by Quine,' 'belongs to some class,' etc. Categories which include no other categories are *types*. While a category (other than a type) can contain two individuals which belong to two other exclusive categories, all the individuals in a type belong to all the same categories. For example, the category of individuals spanned by 'red' is not a type since Mars and I are both spanned by 'red,' but only Mars is spanned by 'has an elliptic orbit' and only I am spanned by 'speaks French.' On the other hand, 'prime number' determines a type since whatever spans any number spans all numbers.

In other words, numbers (unlike red things) all belong to all the same categories and thus form a type. Every individual belongs to one and only one type. While categories cannot overlap, types can neither overlap nor include one another. We might think of them as ultimate classes always exclusive of one another. One way of looking at the issue between dualist and monist theories of persons is by taking it to be a question of whether material objects (things spanned by terms like 'extended,' 'red,' etc.) constitute a type (dualists like Descartes) or merely a category (monists like Strawson).

One recognizes categories (and types) when one sees that spanning is different from truth. Categories divide the things of the world in ways that are ontologically more interesting than those achieved by mere classification.

The Language Tree

What is a language tree? Very simply, a language tree is a map which graphically represents the sense relations among the terms of a language. What is important about the ordinary language tree, according to Sommers, is that it represents, by virtue of the isomorphism of structure between ordinary language and ontology, the ontological structure. Whereas the relations exhibited by the *language* tree are *sense* relations between terms, the relations exhibited on the *ontological* tree are *category inclusion* relations. Since the structures of the two trees are isomorphic, a single map will serve for both. Sommers' task, of course, in expounding this theory is first to construct a feasible language tree and then to establish the requisite isomorpism so that the ontological structure can be "read" from the language tree.

In order to construct a language tree Sommers first sets out the notion of U-relatedness between terms. The theory of meaning which he adopts is the "meaning-in-use theory." Accordingly, one knows the meaning of a given term only if one knows how to use it to make nonabsurd sentences. A complete knowledge of the term would entail the ability to understand every nonabsurd sentence in which that term is used. Since several terms can be interchangeable with respect to such a set of sentences, any map which exhibits such uses of a term cannot give the entire *meaning* of it. What it can give is the *sense*. The sense of a term is just a part (a very important part) of its meaning. "The *sense* of an expression will be its location with respect to other expressions, its semantic range. It is what it 'makes sense with' as contrasted with what it fails to make sense with" ("The Ordinary Language Tree," p. 161). To determine the sense of a term, it is necessary to determine the set of terms which is such that it "makes sense" to use the given term with each member of the set. Any two terms which are such that it makes sense to use them in a subject-predicate sentence (where one is the subject, the other the predicate) are said to have the sense relation *U*. Such terms are "U-related" A *U-set* for a given term will consist of all those terms to which it is U-related. The U-set of X is symbolized as "U(X)." To indicate that two terms (say X and Y) are U-related we will write "U(XY)." Two terms are said to be *N-related* just in case they are not U-related. In other words, if two terms do not "make sense" together they are N-related. Any subject-predicate sentence formed by an N-related pair will be called a "category mistake." Given these definitions we can write the following U and N examples: U(red, house), U(interesting, red), U(interesting, number), U(interesting, person), U(prime, number), U(person, thoughtful), U(person, red), N(red, number), N(prime, house), and N(red, thoughtful). Since the U- and N-relations are obviously symmetrical, the order in which the related terms occur is irrelevant.

Two terms are said to be "connected" just in case they are terms in the same language. Formally the "connectedness conditions" are:

(i) U(X,Y) and U(X,Z) implies *con*(Y,Z)

(ii) *con*(Y,Z) and *con*(W,Z) implies *con*(Y,W)

Thus, while "red" and "thoughtful" are N-related, they are nevertheless connected since they are both U-related to a third term (e.g., "person"). Two terms are part of the same language if and only if they are connected. The set of terms of a language can be defined as the largest set of terms such that each term in the set is connected to every other term in the set.

A model language tree can be constructed by writing all the terms of the language such that a solid line is placed only between U-related terms. Here is an example of a partial tree:

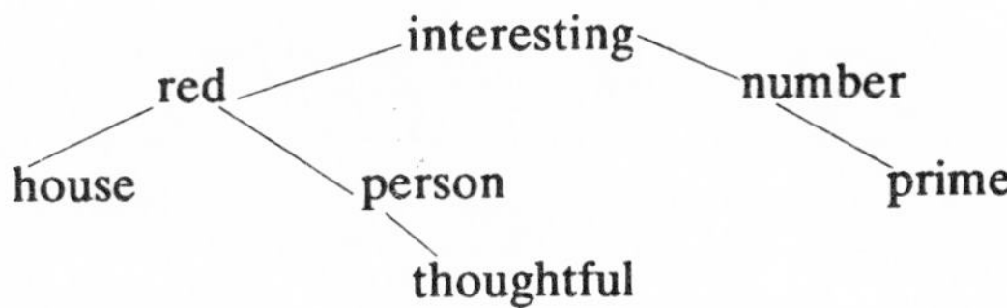

Two terms are U-related if we can go from one to the other by following a continuous downward or upward path. Thus given our earlier U- and N-relations and our mapping of them, we can now extract other U-related pairs such as U(interesting, house), U(interesting, thoughtful), U(interesting, prime), U(red, thoughtful) and N-related pairs such as N(red, prime), N(prime, person), N(prime, thoughtful), N(house, thoughtful), N(number, thoughtful), and N(number, person).

If this is our tree

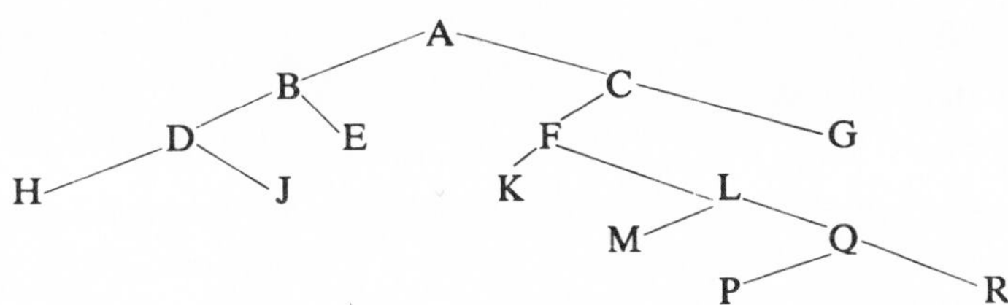

we can see that every term is U-related to A and, for example, Q and D are N-related. If there were no term A, we would simply have two different languages since no member of the set (BEDHJ) would be connected to any member of the set (CFGKLMQPR).

The ontological importance of the ordinary language tree is that it shares a common structure with the ontological tree and so the same tree can be used to "read off" either linguistic *or* ontological information. In "Types and Ontology" one of Sommers' main tasks is to establish a principle of isomorphism between language and ontology. Preliminary to an understanding of this task is an understanding of several elementary principles which Sommers develops in "Types and Ontology."

One of the things Sommers is anxious to argue against is the adoption of what he calls the "transitivity rule" for sense relations. The transitivity rule is the traditional view that if two terms are such that they both "make sense" with some third term then they must "make sense" with each other. We can state the rule like this:

Transitivity Rule: for any three terms, P, Q, R,

$$U(PQ) \quad . \quad U(QR) \quad . \supset U(PR)$$

Sommers' attack on the transitivity rule proceeds as follows. He first distinguishes between four kinds of types, called α-, B-, β-, and A-types. He gives semantical definitions for each in such a way that they are clearly exclusive of one another. Briefly, what Sommers wants to show is that if the transitivity rule holds for U-relatedness, the distinctions between these four kinds of types are obliterated. He gives syntactical definitions for B-types and A-types and then argues that if two terms are U-related they belong to the same B-type and if they are of the same B-type they are U-related. On the other hand, two terms may be U-related yet perhaps not belong to the same A-type. However, if the transitivity of U-relatedness holds, then two terms are U-related just in case they belong to the same A-type. Thus, B- and A-types would be indistinguishable and the earlier fourfold classification of types would disintegrate. So, Sommers concludes that the transitivity rule for U-relatedness must be dealt a "decisive blow," or at least be replaced, which is the task of most of the remainder of "Types and Ontology."

Let us call the complement of a term its "logical contrary." The logical contrary of a term, P, is $\overline{P}$ (read "nonP" or "unP"). The logical contrary of a term will be equivalent to a disjunction of all of the contraries (nonlogical) of the term. For example, the term "red" has as its logical contrary "red." "Red" is equivalent to "blue or green or white or yellow or orange or...," where "blue," "green," etc., are nonlogical contraries of "red." Now, since any term and its logical contrary span the same things we can ignore the value of the term (i.e., whether it or its logical contrary are involved) and consider what Sommers calls the "absolute" term. Absolute P will be written '/P/' and read as the disjunction of P with its logical contrary. As we will see, for the purposes of ontology we need only consider a language tree of absolute terms.

Let us absolutize the predicates which we used earlier for our sample language tree. We then have:

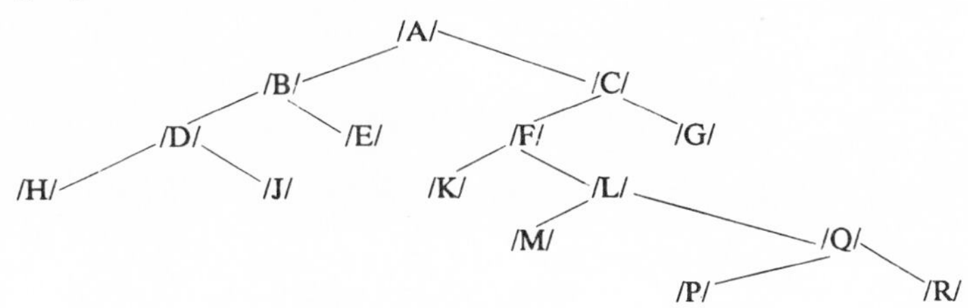

Of course, two or more terms may have the same use (e.g., "red" and "blue") so that we might have several terms at any given "node" of the tree. If /N/ occurs at the same node as /F/ we will say that N and F are "categorially synonymous." Given Sommers' definitions of A- and B-types, we can see that B-type will consist of those terms which are connected to each other by a continuous upward or downward line (i.e., those which are mutually U-related). We have the following B-types: (ABDH), (ABDJ), (ABE), (ACFK), (ACG), (ACFLM), (ACFLQP), and (ACFLQR). Each set of terms at a given node will constitute an A-type since all terms which are categorially synonymous with one another have the same U-sets.

While the A- and B-types are classes of predicates (things in our language), the α- and β-types are classes of things (things in our ontology). Here is the definition of an α-type: a set of things constitute an α-type with respect to a given predicate P if and only if P spans all of the things in the set and nothing outside the set. Since P spans whatever $\overline{P}$ spans, and vice versa, and the set spanned by /P/ is the union of the sets spanned by P and $\overline{P}$, we can substitute /P/ for P in our definition. Then, corresponding to each absolute predicate on the tree there will be an α-type with respect to that predicate. Since categorically synonymous terms span the same things, we can say that for each node on the language tree there exists exactly one corresponding α-type, or category. In other words, given a tree such as ours, we may read the expression "/P/" either as "absolute P" or as "the category with respect to P."

Every member of a β-type is spanned by all and only all the members of a given B-type. The B-types of a tree can be found by tracing a continuously downward path from the top node of our tree to each bottom node, there will be a β-type corresponding to each bottom node on the tree. Bottom nodes are simply special kinds of nodes just as β-types (types) are simply special kinds of α-types (categories).

In effect, then, given any tree such as our sample tree, we can read the symbols on the tree (e.g., "/A/", "/B/", "/C/", etc.) either as absolute predicates or as categories. So, our tree when "read" in one way is a map of sense relations between terms, and when "read" in the other way is a map of relations between categories. The task of establishing an isomorphism between the linguistic and ontological structures appears achieved.

However, an important question is yet to be answered before this is the case. The question is: What kind of relation between categories is represented on the ontological tree? When the tree is "read" as a language tree, the lines between terms represent U-relatedness. What do these lines represent when the tree is "read" as an ontological tree? In order to establish a complete isomorphism between language and ontology Sommers had to find some principle which allowed him to "translate" U-relatedness into some relation between categories. This principle is the requisite principle of isomorphism between language and ontology. "Indeed the law of categories and the law governing the distribution of category mistakes (N-related pairs) are two expressions of a structural isomorphism which holds between 'language' and 'ontology.' In its categorial form the structural principle may be thus stated:

> If C_1 and C_2 are any two categories, then either C_1 and C_2 have no members in common or C_1 is included in C_2 or C_2 is included in C_1 ("Types and Ontology," p. 355).

Sommers calls this the "Law of Category Inclusion." His answer to our question is, in effect, that corresponding to the U-relation on the language tree is the category inclusion relation on the ontological tree. What is more, one version of this law is just the replacement required for the Transitivity Rule.

Rules

Some new rule had to be found which would retain transitivity of type (β-type) sameness, enforce only necessary ambiguity of terms and yet avoid the faults of the transitivity rule. Included in Sommers' program is an attempt to show that "linguistic structures and ontological structures are isomorphic." Such a structural principle is found by looking to the categories that are of fundamental importance for ontologists. "There are two sorts of categories which are of major importance to ontologists. These are the categories which are all inclusive, containing all others as subcategories, and those that are completely exclusive, containing no subcategories at all" ("Types and Ontology," p. 354). If there is such an all-inclusive category, then the predicates which define it (i.e., the terms at the apex node of the language tree, "apex" terms) must be U-related to every other term in the language. If there are such all-exclusive categories, then the predicates which define them (i.e., the terms at the bottom nodes of the language tree, "bottom" terms) must be U-related to a finite set of other terms, all of which are mutually U-related. In other words, such bottom terms must belong to only one B-type.

The structural principle which Sommers offers is the Law of Category Inclusion. This law, along with its derivative theorems, satisfies equally well Sommers' requirement for a law which establishes isomorphism between linguistic and ontological structures, and his requirement for a replacement for the transitivity rule. He initially presents the law unformalized.

> The existence of dominating categories and categories of the lowest level is however assured by a fundamental law which governs all categories. This law can be derived from a 'syntactical' rule governing the distribution of category correct statements in a natural language. Equally, the rule can be derived from the law. Indeed the law of categories and the law governing the distribution of category mistakes are two expressions of a structural isomorphism which holds between 'language' and 'ontology.' In its categorical form the structural principle may be thus expressed:
>
> > If C_1 and C_2 are any two categories, then either C_1 and C_2 have no members in common or C_1 is included in C_2 or C_2 is included in C_1.
>
> Given this law and the already noted fact that the categories...defined by the predicates of any natural language are finite in number, it will follow that there must be one category that includes all others and several that include no others.
>
> I shall call this the law of categorial inclusion since it states that whenever two categories have some common membership one of the two is included in the other ("Type and Ontology," p. 355).

I have quoted at length because several points made in this passage are extremely important for the development of Sommers' program. Here also is a convenient place to point out that Sommers is correct in saying that given this law and the fact that the categories defined by a natural language are finite, it follows that there must be one all-inclusive category and several all-exclusive categories. This simply means that since the set of categories is finite, there must be both a lower and an upper limit to the number of possible inclusion combinations between any two members of the set.

Clearly the main body of Sommers' theory rests upon the acceptance or rejection of the law of categorial inclusion along with its derivative theorems. Sommers gives three theorems in "Types and Ontology."

T.1, the law of categorial inclusion, is formalized by Sommers in three equivalent ways:

$$\text{(T.1)}\quad U(PQ)\equiv((/P/\subset/Q/)\ v\ (/Q/\subset/P/))$$
$$U(PQ)\equiv(x)\ (/P/x\supset/Q/x)\ v\ (x)\ (/Q/x\supset/P/x)$$
$$U(PQ)\equiv(x)((x\in/P/)\supset(x\in/Q/))\ v\ (x)((x\in/Q/)\supset(x\in/P/))$$

Here Sommers has justifiably used the symbol "/P/" for both the absolute predicate and for the category which it defines. Also "v" is used inclusively, and '⊂' is used for simple (rather than proper) inclusion.

The first theorem purportedly derivable from T.1 holds for any P, Q, and R:

$$\text{(T.2)}\quad U(PQ)U(QR)N(PR)\equiv(/P/\subset/Q/)(/R/\subset/Q/)(\overline{/Q/\subset/P/})$$
$$(\overline{/Q/\subset/R/})(\overline{/P/\subset/R/})(\overline{/R/\subset/P/})$$

> The importance of T.2 lies in its clear statement of an equivalence between syntactical and categorical relations, between sense and nonsense gotten by conjoining terms, and the inclusions among the sets of things they span. From another side, T.2 states a simple criterion which validates the subject-predicate distinction. To see this clearly, let us substitute the symbol "←" wherever we have the inclusion symbol "⊂" and let us interpret "Q←P" to mean: "of (what is) P it is significant to say that it is Q." Or--what is the same thing--"that it is Q" is predicable of (what is) P. We now have
>
> $$U(XY)U(XZ)N(YZ)\equiv(X\leftarrow Y)(X\leftarrow Z)(\overline{Y\leftarrow X})$$
> $$(\overline{Z\leftarrow X})(\overline{Y\leftarrow Z})(\overline{Z\leftarrow Y})$$
>
> This tells us that the middle term X of two significant pairs (sentences) is always the predicate with respect to the two N-related terms Y and Z of a triad, U(XY)U(XZ)N(YZ). The relational expression "is predicable of" is seen to be isomorphic with the rational expression "contains." Both the relations of predication and of containment are transitive and nonsymmetrical. In this respect the predicative tie between terms differs from the tie of significance of U-relatedness since the U-relation is, as we have seen, nontransitive and symmetrical ("Types and Ontology," p. 356).

Thus the law of categorial inclusion, by virtue of T.2, is a replacement for the transitivity rule in so far as it retains transitivity for type sameness while rejecting transitivity for U-relatedness. Moreover, the relation shows that predication is nonsymmetrical and, according to Sommers, it demonstrates the need for the traditional subject-predicate distinction and serves as a test for finding which term of a pair is the predicate. Obviously T.2 is a key theorem in Sommers' program.

The law of categorial inclusion, along with its two derivative theorems, is meant to serve not only as an expression of structural isomorphism between language and ontology, but also as a replacement for the transitivity rule. We have seen that the law of categorial inclusion, by virtue of T.2, is meant to be a replacement for the transitivity rule insofar as it retains transitivity for type sameness while rejecting transitivity for U-relatedness. But, for the law to serve as a complete replacement for the transitivity rule, it must be able to be used to enforce only necessary ambiguity. This is part of the role fulfilled by the second derivative theorem.

(T.3) –(U(PQ)U(QR)U(PS)N(PR)N(QS))

Now, what T.3 says in effect is that any term which is U-related to two N-related terms is U-related to any term which is U-related to either of those terms. In other words, any term Q which is U-related to two N-related terms, P and R, is also U-related to any term U-related to P or R. So if S is U-related to P, then Q and S cannot be N-related, i.e., –(U(PQ)U(QR)U(PS)N(PR)N(QS)). This rule, which Sommers calls the "tree rule" since it allows one to locate any term on a language tree, is certainly worth preserving. I am hard pressed to find any four terms which could possibly constitute a counter-example to this rule. As a matter of fact, if we look at the tree for the values U(PQ)U(QR)U(PS)N(PR)N(QS), we find that they cannot all occur simultaneously. Since U(PQ), U(QR), and N(PR), the tree looks first of all like this:

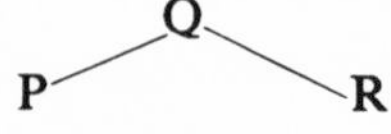

But now if also U(PS), then either

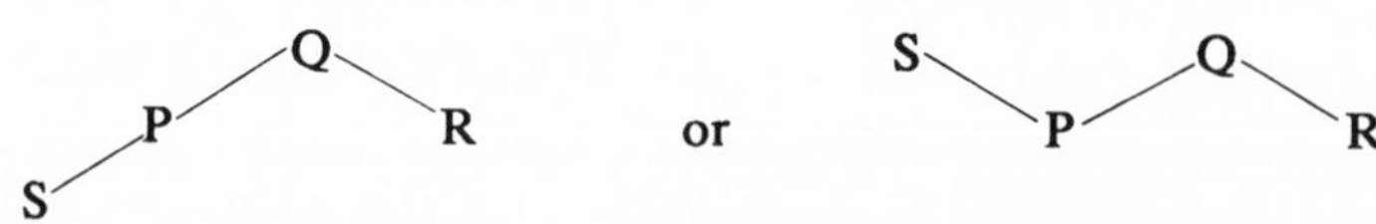

If one can go from one term to another by following a continuously downward or upward path, then the two terms are U-related. Since N(QS), the first tree cannot be correct, leaving the second tree as the only possible one for the given values. However, Sommers has said in his informal introduction of T.1 that if two categories have common membership one must be included in the other. Now, looking at the second tree, the law of categorial inclusion tells us that either (i) /S/ and /Q/ are included in /P/ or (ii) /P/ is included in /S/ and in /Q/. But (i) cannot hold since if /Q/ is included in /P/, and /R/ is included in /Q/, /R/ is included in /P/, which is impossible since N(PR); and if /Q/ is included in /P/, and /Q/ is included in /R/, then /Q/ constitutes a common membership for /P/ and /R/ such that one must be included in the other, which again is impossible since N(PR). Moreover, (ii) cannot hold since if /P/ is included in both /Q/ and /S/, it constitutes a common membership for /Q/ and /S/ so that either /Q/ is included in /S/ or /S/ is included in /Q/, which is likewise impossible since N(QS). Thus, there seems to be good reason for wanting T.3 as a rule and showing that it can be derived from the law of categorial inclusion. T.2 tells one how to translate U(PQ)U(QR)N(PR) into a statement about category inclusions. If again we construct a tree for these three values, we obtain:

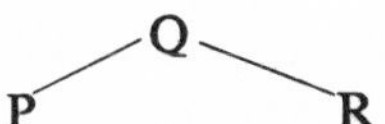

The important thing about T.2 is that it denies that in the given case /Q/ is included in either /P/ or /R/, i.e., ($\overline{/Q/\subset/P/}$) and ($\overline{/Q/\subset/R/}$). Obviously, if /Q/ were included in /P/ and in /R/, then /P/ and /R/ would share some common membership, making ($/P/\subset/R/$) or ($/R/\subset/P/$). Any two terms which are N-related to each other but U-related to a third term must both define categories which are included in the category defined by the third term. If this third category were included in either one or both of the first two, then the first two terms would be U-related. In fact, this is exactly what T.2 says;

$U(PQ)U(QR)N(PR) \equiv (/P/ \subset /Q/)(/R/ \subset /Q/)(\overline{/Q/ \subset /P/})(\overline{/Q/ \subset /R/})(\overline{/P/ \subset /R/})$ $(\overline{/R/ \subset /P/})$. So, as with T.3, there is good reason for holding T.2 as a rule.

Finally, we could add to these three rules:

$$\text{(T.4)} \quad (P)(\exists x)(x \in /P/)$$

Sommers suggests this rule in "Predicability" (p. 277). What it says is that no category is empty.

We have seen that the main feature which the tree theory displays is an isomorphism between language and ontology. Sommers' theory is simply the fulfillment of Russell's program, which held that "(a) clarification of natural language is ontologically revealing and discriminatory of the sorts of things there are; (b) linguistic structures and ontological structures are isomorphic ("Types and Ontology," pp. 350-351). In effect, what the tree theory says is that certain pieces of nongrammatical linguistic information (i.e., facts of language independent of any particular natural language) parallel some pieces of ontological information. Thus, the business of doing ontology can be gotten at either by looking to the linguistic structure or to the ontological structure.

This isomorphism between language and ontology is complete. For every term in the language there is a corresponding category. For terms that are categorially synonymous (i.e., occuring at the same node on the language tree) the categories which they define are identical. Thus, given a tree, we can read the symbols on it either as terms or as names of categories. Moreover, T.1 guarantees that the expression

$$P \text{———} Q$$

found on a tree can be read either as an expression of a U-relation between two terms or as an expression of a category inclusion relation between two categories. And in Sommers' discussion of T.2 we learn that the relation "is predicable of" is isomorphic with the expression "includes." In other words, given any tree, we can interpret it either as a language tree or as an ontological tree.

Rules which express only relations between terms, we might call *sense* rules. T.3 is an example of a sense rule. Corresponding to each sense rule will be a rule expressing only relations between categories (or the members of categories). We might call such a rule a *categorial* rule. The law of categorial inclusion is such a rule. The categorial rule parallel to T.3 might be Sommers rule about "category straddlers." "Strictly speaking there are no category straddlers" ("Predicability," pp. 279-280). In other words: it is not the case that there is a thing which belongs to two categories neither of which include the other. Hence, given a tree, if we consider it as a language tree, we can read off T.3. If we consider it as an ontological tree we can read off the rule about category straddlers. They tell us, in effect, that no tree can contain

whether S, Q, P, and R are terms on the one hand or categories on the other.[7]

Besides sense rules and category rules, there are rules which actually express the isomorphism between language and ontology. The reason we can "translate" back and forth between language and ontology is because any symbol on the tree can be taken either as a term or as a category, and every pair of symbols connected by a continuously upward or downward line can be taken as expressing either a U-relation or a category inclusion. T.1 and the definition of a category (α-type) are such rules. Rules like these give not sense information, like T.3, and not categorial information, like the straddler rule, but rather define a connection between terms and categories or between term relations and category relations. Such rules might appropriately be called *translation* rules. A translation rule tells us something about both a language and an ontology.

Sommers' "rule for enforcing ambiguity" is the most important translation rule.

> If *a*, *b*, and *c*, are any three things and *P* and *Q* are predicates such that it makes sense to predicate *P* of *a* and of *b* but not of *c* and it makes sense to predicate *Q* of *b* and of *c* but not of *a*, then *P* must be equivocal over

> *a* and *b* or *Q* must be equivocal over *b* and *c*. Conversely, if *P* and *Q* are univocal predicates, then there can be no three things such that *P* applies to *a* and *b* but not to *c* while *Q* applies to *b* and *c* but not to *a* ("Predicability," pp. 265-266).

Sommers' commentators and critics have misunderstood this rule on one of two grounds. Either they have failed to see what kind of rule it is or they have failed to recognize completely its force. In the first group[8] are those who take the rule as a sense rule rather than a translation rule. They mistakenly believe that the rule is a test for discovering term ambiguity. As a clear matter of fact, the rule is no such thing. Rather than being a test for term ambiguity, it is a test for theory coherency. "Note that the rule for ambiguity does not tell us *which* of two predicates is ambiguous...we apply such a rule to check the coherency of philosophical positions..." ("Predicability," p. 266).

In the second group[9] are those who, while seeing that the rule is one for *enforcing* ambiguity rather than *discovering* it, still fail to recognize the full force of the rule as a tool for testing the coherency of a theory. This is the result of their failure to count the last sentence of Sommers' statement of the rule as a necessary part of the rule. This sentence says, "Conversely, if *P* and *Q* are univocal predicates, then there can be no three things *a*, *b*, and *c* such that *P* applies to *a* and *b* but not to *c* while *Q* applies to *b* and *c* but not to *a*." In other words, confronted with the task of rendering coherent a theory which allows two predicates, *P* and *Q*, and three things, *a*, *b*, and *c*, standing in the requisite relations, our choice is not simply to make one or both of the predicates ambiguous. Our choice is much greater. We can (1) make *P* ambiguous, or (2) make *Q* ambiguous, or (3) deny the existence of *a*, or (4) deny the existence of *b*, or (5) deny the existence of *c*. Thus, instead of having just three choices ((1), (2), or (1) and (2)), we have thirty choices (the possible combinations of (1) through (5)). At this point, to deny the existence of a thing is simply to deny that it is in the ontology (i.e., to deny that it is an individual). To say that Descartes denies the existence of heavy thinking Jones is to say that Descartes denies that Jones is an individual.

In summary, for those who wish to use Sommers' rule for enforcing ambiguity, the following points must be kept in mind: (a) the rule is not a sense rule for testing term ambiguity, (b) it is a translation rule for testing theory coherency, and (c) the enforcement of coherency upon a theory which has two predicates and three things in the described relations need not be the result just of enforcing ambiguity on one or both of the predicates, but rather, may result from any number of other possible ways of applying the rule such as denying existence to one or more of the three things. The rule for enforcing ambiguity is the heart of Sommers' programme.

We might call all of the rules which are the concern of the ontologist *ontological* rules, whether they be sense rules, category rules, or translation rules. An ontologist can get at the business of doing ontology by looking either to the sense rules or to the category rules because they are "isomorphic." But as we have seen, if one of his tasks is to ensure theory coherency, he does this best by using the translation rules such as the rule for enforcing ambiguity.

Actually, the business of the ontologist is, in a way, threefold. His concern with language is directed at avoiding nonsense (i.e., category mistakes). His concern with ontology is directed at avoiding non-individuals (i.e., category straddlers). These two collapse into his general concern which is directed at avoiding theory incoherency. Thus, in judging a theory to be incoherent by use of the rule for enforcing ambiguity, the ontologist's verdict can be taken in either of two ways. Either the guilty theory can be said, in one mode, to allow some category mistake to be true, or, in the other mode, to allow some category straddler into its ontology. A category straddler, or non-individual might be thought of as a category mistake in the material (ontological) mode.

Ontology

A thing is categorially possible just in case it is an individual (i.e., all the terms which span it are mutually U-related). We are also able to say what it means to say that some kind of class of things is categorially possible. Let us say that D-things (things of which D is true) are categorially possible if and only if D spans only individuals. More formally,

D-things are categorially possible ≡df
$(x)(x\in/D/\supset.(P)(Q)(x\in/P/\cdot x\in/Q/.\supset U(PQ))$[10]

DC-things will be categorially possible just in case U(DC). Obviously, then, odd even things are categorially possible but not red even things. In general, $D_1...D_n$ things are categorially possible just in case $D_1...D_n$ are all mutually U-related.

$(D_1...D_n)$-things are categorially possible ≡df $U(D_1...D_n)$ In short, for any D, whether D is a simple term or a compound term, D-things are categorially possible if and only if D is a term on the language tree, since only terms which span only individuals occur on the tree.

Talk about things being categorially possible, then, is really talk about predicability relations among terms. As we have seen, "U(PQ)" can be defined as "P is predicable of Q or Q is predicable of P," which further reduces to "P spans whatever Q spans or Q spans whatever P spans." We have, for example, DC-things are categorially possible just in case D spans whatever C spans or C spans whatever D spans.

An obvious objection might be raised now. When D is equivalent to an N-related pair (e.g., "red number"), then DC-things are categorially possible since C would span at least whatever D spans, viz. nothing. We would be forced to say that for any ϕ, red numbers which are ϕ are categorially possible, which is obvious nonsense. But are terms equivalent to N-related sets of terms U-related to every term?

Let us say that a term is whatever can serve as the subject or predicate of a grammatically correct sentence. A predicable term is a term which belongs to the set of mutually connected terms of a language. Such a term will thus be U-related to at least one other term. An impredicable term is any term which is not a predicable term. "Red" is a predicable term, "number" is a predicable term, but "red number" is not a predicable term. Things spanned by such impredicable terms are non-individuals (i.e., not in the ontology in the same way that such impredicable terms are not in the language). Only U-related terms can be used to form new (predicable) terms. As we have seen, recalling T.1p, every term (predicable or not) spans something (thus we can reply directly to our critic above that it just is not the case that terms like D, "red number," span nothing; they at least span a thing. Every predicable term spans some individual. No impredicable term spans any individual. The question of whether or not impredicable terms are U-related to every term arises only when we mistakenly view such terms as determining empty categories, which are thus included in all other categories. But, given T.4, there are no empty categories. Even an impredicable term spans some thing.

We can now speak more clearly and rigorously about the notions of *thing* and *individual*. We will say that x is a thing if and only if some term spans x. Thus red numbers are things since they are spanned, at least, by "red numbers." We will say that x is an individual if and only if any pair of terms spanning x are U-related. Since only predicable terms can ever be U-related and since only U-related terms can form new predicable terms, this amounts to saying that x is an individual if and only if whatever spans x is a predicable term. Since individuals are things, if x is an individual whatever spans x is a predicable term and some term does span x. It follows that whatever is an individual is spanned by some predicable term.

Red numbers and red houses are things since some term spans each. But, red houses are individuals while red numbers are not. All the terms spanning red houses are mutually U-related and thus no impredicable term spans red houses. On the other hand, red numbers are spanned by at least two terms not U-related ("red" and "number"). Since any term formed by two predicable but N-related terms is itself an impredicable term, red numbers are spanned by at least one impredicable term ("red number") and thus are non-individuals.

We might very well ask what kind of ontological commitment must be made by a speaker. The answer is clear. He is committed to just those things which constitute his ontology, i.e., just those things which are categorially possible, individuals. Since what is taken as an individual is determined by the kinds of predications one is willing to make (indeed, restricted to making), on Sommers' view ontological commitment is determined not by what one is willing to refer to, not by what one is willing to quantify over, but by what predications one is willing to make. Indeed, we might say that our ontological commitment is made when we establish (through use) the U- and N-relations among terms. Those of us who follow the same sense rules when using language have the same ontological commitment (are committed to the same ontology). When we refrain from allowing color terms to span numbers when we talk, then we are committed to an ontology free of colored numbers. When we are willing in our talk to say of circles that they fail to be square (allow "square" to span circles), then we are committed to an ontology which includes square circles as well as nonsquare circles. The categories of things which constitute our ontology change only when our way of talking (i.e., our ways of predicating, our U- and N-relations) changes. Clearly, my ontology does not tell me very much about what there is. Of course, there are nonsquare circles; but *square* circles! But ontology is only meant to provide *ontological* information. It takes us only so far. In fact, it takes us only up to the ontological level, which is admittedly not very high. It happens to be lower even than the logical level. What good then is ontology for telling us what there is? It might tell us that our ontology contains both square and nonsquare circles, but it cannot tell us which are logically possible and which are not, and it certainly cannot tell us which exist. It can tell us, say, that our ontology allows persons but not their minds, but it cannot tell us whether there are such things. It can tell us that odors and colors rather than skunks are found in our ontology, but it cannot tell us about the existence of such things. It can discern both Hamlet and Shakespeare in our ontology, but does not provide a clue as to which exists and which does not. The point is that while ontology does not take us very far in deciding about what there is (i.e., what one is existentially committed to), it is an essential first step.

Obviously whatever one is committed to existentially, he is committed to ontologically. However, it certainly need not be the case that one is existentially committed to whatever he is ontologically committed to. I am ontologically committed to both square and nonsquare circles, but I am, of course, not existentially committed to both.

Let us consider the term "exists." Since no statement of the form *a exists* is a category mistake it seems safe to say that *exists* is an apex term (U-related to every term in the language). In our ontology (the one isomorphic with our ordinary language), then, everything belongs to the *category* of existence. Every individual either exists or fails to exist (note that nonindividuals, things outside the ontology, neither exist nor fail to exist). What of those philosophers, like Aristotle, who argue that "exists" is ambiguous and thus not necessarily at the apex of the language tree? Sommers has said that a term is univocal until the observance of rules forces ambiguity upon it ("Types and Ontology," p. 350). Another look at the rule for enforcing ambiguity shows that ambiguity can only be forced upon a term which defines a category which shares some membership with a second category but neither includes nor is included in that category. Another way of putting this is: ambiguity can only be forced upon a term which co-spans an individual with another term neither higher than, nor lower than, nor categorially synonymous with it (i.e., N-related to it). Clearly, then, ambiguity could never be forced upon any genuinely apex term, since every term is either lower than or categorially synonymous with any apex term (i.e., is U-related to it). If "exists" spans every individual, then it is an apex term. If "exists" is an apex term, then it is univocal.

We might say, then, that ontology tells us what a theory says there /is/ (what /exists/). We are ontologically committed to whatever belongs to the *category* of existence, i.e., the set of things *spanned* by "exist." This category includes the *class* of things which a theory holds to exist and the *class* of things which it maintains fails to exist.

Obviously we do not take everything which /exists/ to exist. After discovering what /exists/ (i.e., the ontological commitment of a theory) our next task is to find a way of eliminating those individuals which the theory excludes from existence (the ontology has already excluded those things which neither exist nor fail to exist, nonindividuals). Once we have found a way of doing this, then what remains will be what is taken to exist. Just as we are ontologically committed to just those things which constitute our ontology (viz. the *category* of existence), we will be existentially committed to just those things which we allow to constitute the *class* of existence.[11]

How, then, are we to discern between what a theory says exists and what it says fails to exist? Let us say that *nothing impossible exists*. In other words, whatever is held as impossible, in any sense, is excluded from the existential commitment of a theory. At the ontological level we have already ruled out categorially impossible things (e.g., red numbers and valid philosophers). Note that what is excluded from one's existential commitment may or may not fail to exist, i.e., may or may not also be excluded from his ontology. At this level we are left with only our ontology, individuals, things which /exist/. At the next level, the logical level, we are able to eliminate those things which we are committed to ontologically, but which we nevertheless take to be impossible (logically). If we are ontologically committed to whatever is not eliminated at the ontological level, then we are *logically* committed to whatever is not eliminated at the logical level.

Our logical commitment is clearly included in our ontological commitment, but it cannot be identical with our existential commitment since it includes things which are logically possible but nevertheless obviously fail to exist (e.g., dog sleds faster than light and pregnant mules).

Things such as these must be ruled out at the empirical level. While, ordinarily, we are ontologically committed to square circles and pregnant mules, and logically committed to pregnant mules, we need not be empirically committed to pregnant mules. We are empirically committed to just those things not eliminated at the empirical level. We have seen that our ontological commitment *properly* includes our logical commitment, which in turn *properly* includes our empirical commitment. The question which remains now is this. Is our existential commitment properly included in our empirical commitment or is it identical with it? The first alternative seems to be the obvious answer. We are empirically committed to just those logically possible things which are also empirically possible. But, not everything which is empirically possible need be empirically actual (exist). The class of empirically possible things (what one is empirically committed to) is the union of two classes: (i) those things which are taken as possible in every sense but which one still holds fail to exist, and (ii) those things which are not only taken as possible in every sense but are held to be actual as well, i.e., those things to which one is existentially committed. For example, for the ordinary speaker, the present president and the present king of France are both possible in every relevant sense, but only one exists.

What are we existentially committed to, then? Just those things which we are willing to say exist. Such things will be individuals, logically possible, empirically possible, and actual.

The following diagram might serve to illustrate the notion of levels of rectitude involved so far.

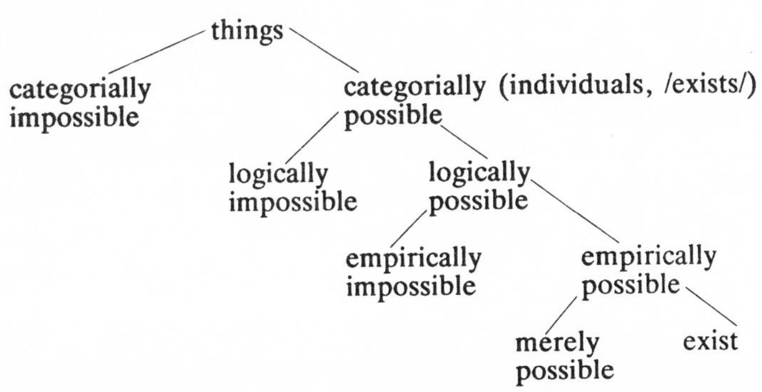

From this we can see that the things which a speaker or theory takes to exist are empirically, logically, categorially possible things. What it excludes from the class of existence are things which are impossible in any sense or *merely* possible in any sense. What determines our ontological commitment (ontology), or logical commitment, or empirical commitment are the linguistic rules (sense, logical, laws of science) which we apply at those respective levels. On the other hand, we cannot distinguish between what exists and what is merely empirically possible by looking at these kinds of rules. We can say at least this much for now. Ontological, logical and empirical commitments are all presupposed by existential commitment. The first step in discerning an existential commitment (i.e., "what a given remark...says there is..."[12]) is a recognition of the differences between the *category* and the *class* of existence. The ontologist's concern is with categories rather than classes. The limits of a theory's ontology are determined by what it says /exists/, not by what it says exists. To paraphrase Wittgenstein's remark: The limits of my language are the limits on my ontology.

Concluding Remarks

James W. Cornman said, "That no ontological position follows from the work of the logical analyst can be seen from the fact that whereas the logical analyst is solely concerned with the relationships among various linguistic expressions (e.g., he is concerned with paraphrasing sentences or showing how the functions of certain expressions are different from or similar to the functions of other expressions), what is necessary for ontology are the relationships between linguistic expressions and entities."[13]

I hope that what I have done in this chapter shows that Cornman's remark was wrong. It is indeed the case that "what is necessary for ontology are the relationships between linguistic expressions and entities." But, as Sommers has clearly shown, it is not at all the case that "the logical analyst is solely concerned with the relationships among various linguistic expressions." A very definite ontological position follows from the work of such a philosopher once he is able to establish a principle of isomorphism between language and ontology.

This principle, expressed in the categorical form, is the law of category inclusion. Sommers' entire ontological theory is based upon this law and his definition of a category in terms of *spanning*. The importance of this law for ontology cannot be overemphasized. Expressed as a translation rule it becomes the rule for enforcing ambiguity. It is the application of this rule which distinguishes coherent from incoherent theories. Those who fail to see the importance of the law must fail to appreciate the coherence requirement for theories. Those who wish to reject Sommers' tree theory must eventually reject the law of category inclusion. Yet, it seems to me, that to reject this law is to deny certain fundamental, obvious facts. Indeed, to reject the law is to allow category straddlers, which, in another mode, is to allow category mistakes. Those who wish to avoid category mistakes must accept Sommers' law.

One must keep in mind that Sommers has not *proposed* that we look at ontology and language in this way. He has simply accounted for those facts true of how we talk which explain why and how we ordinarily avoid category mistakes. As Ryle pointed out, it is the category mistake which is the unique property of the philosopher. While admittedly the avoidance of category mistakes is not the end of philosophy, it may very well be the beginning.

Chapter I

Endnotes

* A shorter version of this chapter first appeared in French as Englebretsen (1988a).

1. See Sommers (1959), (1963), (1963a), (1964), (l965a), (1966), (1966a), (1970-71), (1971) and (l978).

2. Brody (1972), Chandler (1968), Cogan (1976), Cornman (1968), DeSousa (1966), Drange (1966), Elgood (1970) and (1974), Englebretsen (1971), (197la), (1971b), (1972), (1972a), (1972b), (1972d), (1974), (1974b), (1975), (1975c) and (1976), Erwin (1970), Fjeld (1974), Greenburg (1972), Goldstick (1974), Guerry (1967), Haack (1968), Keil (1979), Lappin (1981), Martin (1969), Massie (1967), Nelson (1964), Odegard (1964) and (1966), Passell (1969), Reinhardt (1965-66), Richmond (1971) and (1975), Sayward (1976) and (1981), Sayward and Voss (1972) and (1980), Swiggart (1972), Suzman (1972), Swanson (1967), VanStraaten (1968) and (1971) and Watson (1973).

3. *The Monist*, 66 (1983).

4. Sommers' most important Works in this area are (1967), (1969), (1973a), (1976), (1976a), (1978), (1981), (1983b), and (1987).

5. Sommers (1982).

6. The distinction is explored in Sommers (1963a) and (1965a) and in several of Englebretsen's works. See also Englebretsen (1981).

7. Actually the straddler rule and the law of category inclusion amount to the same thing. At any rate, T.3 and the law of category inclusion are "isomorphic" (i.e., the same rule expressed in one case as a sense rule and in the other as a category rule). It is this kind of isomorphism between sense and category rules that Sommers refers to in establishing a language-ontology isomorphism. See Sommers (1964a), p. 527.

8. See the papers by Haack and Reinhardt listed in note 2 above. Haack's arguments rest on her view that the rule is a "criterion of equivocity" (p. 159); as such it fails to always tell us that an ambiguous term is ambiguous and often tells us that a univocal term is ambiguous pp. 160-161).

9. See the papers by Chandler and VanStraaten listed in note 2 above. VanStraaten says, "The rule is simply a statement of the logical conditions under which a predicate...must be equivocal"

(p. 61). He ignores completely the force of the last sentence in the rule.

10. (1966) Sommers gave a different definition of "D-things are categorically possible," but has admitted its need for revision. See Guerry (1967).

11. See Englebretsen (1975d).

12. Note that while we have abandoned Quine's criterion for existential commitment our notion of the nature of such a commitment is his. Whichever criterion is used for discerning such a commitment, it tells us not what exists but what a given person or theory says exists. Thus we do not find out what there is (i.e., the existential commitment of a person or theory) by looking at things, but by attending to the uses of terms. For more on existence see Chapter X below.

13. Rorty (1976), p. 166.

Chapter II

SOMMERS ON THE SUBJECT OF A SENTENCE

1. That the grammatical subject of a sentence need not be what that sentence is about was recognized by Aristotle. In *Posterior Analytics* (83a) he compares pairs of sentences such as 'The white (thing) is a log' and 'The log is white.' According to Aristotle, there is an important difference here. In the first sentence the predicate is predicated "improperly." The predicate is "proper" in the second sentence. A proper predication occurs whenever the grammatical subject of a sentence is a substance term. An improper predication occurs whenever the grammatical subject of a sentence is not a substance term, in particular, when it is a quality term (e.g., 'white'). In such cases the predicate is not being predicated of the quality but rather of the underlying substance in which that quality inheres. In the case of 'The white (thing) is a log' the predicate is predicated of what happens to be white. In 'The musician is white' the term 'white' is affirmed of the man who happens to be musical. We might think of Aristotle's improper predications as disguised "proper" predications. The predication of 'log' to 'the white' is like this. It is an apparent improper predication of 'log' to 'the white,' but it is a hidden proper predication of 'log' to 'the thing which happens to be white.' (Notice that 'thing' here must be seen as a variable term for substantives.)

When an improper predication is reformulated as a proper predication the initial, nonsubstantive subject term becomes a predicate term. In proper predications subjects are substantives and predicates are nonsubstantives. After contrasting improper and proper predications Aristotle says, "If we must lay down a rule, let us entitle the latter kind of statement predication, and the former not predication at all, or not strict but accidental predication. 'White' and 'log' will serve as types respectively of predicate and subject."

The problem which arises from constructing sentences with terms like 'white' as their grammatical subjects was particularly acute for Aristotle since in Greek (unlike English) there is little strain on joining the definite article to an adjective, e.g., τὸ μέγα (the large). In English we all feel the strain and immediately and implicitly supply the missing substantive variable, 'thing'--'the large (thing).' The absence of such a strain forced Aristotle to note the hidden substantive explicitly. So, if we cannot invariably determine what a sentence is about by looking to its grammatical subject, how *do* we determine what a sentence is about? At one time Sommers argued that what we are looking for is the *natural* subject of a sentence.[1] While the natural subject of a sentence may be its grammatical subject, it need not be. To see how the natural subject is determined one must look at his tree theory (see chapter I above).

The key structural rule for the language tree is stated most clearly as rule T.3.

(T.3) –[U(PQ)U(QR)U(PS)N(PR)N(QS)]

Suppose we know that U(PQ), U(QR) and N(PR). This is represented on the tree by the following tree fragment.

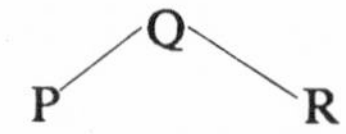

What T.3 says, in effect, is that if, given this fragment, there is a fourth term, S, which is U-related to P it cannot fail to be U-related to Q also. In other words (i) is not possible but (ii) is.

(i) S P Q R

(ii) S P Q R

T.3 is a (indeed, *the*) key structural principle since it requires that a descending tree branch continues to descend and never ascends. A term like Q, which is U-related to terms which are not U-related to each other (P and R) is said to be higher than those terms. T.3 demands that higher here is an asymmetric relation. Thus, it rules out (i), where Q is higher than P and P is higher than Q. Since more than one term can occur at the same tree node (e.g., 'red' and 'yellow'), some pairs of terms are such that neither is higher than the other.

According to Sommers, in a subject-predicate sentence consisting of the terms A and B, A is the natural subject if B is higher than A. Consider Aristotle's 'The white (thing) is a log.' Since there are terms (e.g., 'sky,' 'ink') U-related to 'white' but N-related to 'log,' 'white' is higher than 'log' on the language tree. So 'log' is the natural subject of 'The white (thing) is a log.' Equipollent sentences always share the same natural subject, but need not share the same grammatical subject. What a sentence is about is its natural subject.

2. There is a sense in which Quine's dictum, that to be is to be the value of a bound variable, is a claim that our sentences are about what we take to exist. What we take to exist is revealed by our referential expressions. In a regimented language, such as that of the standard first-order predicate calculus, these expressions are bound variables. The values of our bound variables constitute the "universe of discourse." Thus, what a sentence is about is the universe of discourse with respect to which that sentence is used.

Identifying the universe of discourse for a sentence with the existential commitment made by the use of that sentence has entangled contemporary logic in a mare's-nest of ontological issues (see chapter X below). Nonetheless, there is something intuitively appealing in the notion of a universe of discourse. Sommers has kept the notion as a tool for eluminating what a sentence is about. But he has attempted to free it from the issue of existence.[2] To understand Sommers' notion of a universe of discourse we must return briefly to the tree theory. Sommers argued that any term and its negation have the same sense.

In other words, they share the exact same U- and Nrelations to all other terms. Any term which makes sense with 'red' likewise makes sense with 'nonred.' So one can treat terms on the language tree as neutral, ignoring whether they are positive or negative. Here Sommers introduced the notion of an absolute term, constructing trees from absolute terms such as /P/. The one-to-one correspondence between absolute terms in the language and categories in the ontology allows us to use the /P/-notation for either the absolute term or the category with respect to that term. Sommers' main isomorphism thesis for language and ontology depends upon this term-category correspondence and the existence of a structuring principle for categories corresponding to T.3 for terms. This principle of ontological structure is "the law of category inclusion." The relation between the linguistic structure (determined by T.3) and the ontological structure (determined by the law) is expressed in rule T.1.

(T.1) $U(PQ) \equiv .(/P/ \subseteq /Q/) \vee (/Q/ \subseteq /P/)$

In an appendix to "Types and Ontology" Sommers says that a sentence is about its universe of discourse, and that the universe of discourse for a sentence is the intersection of the categories determined by its terms. If the terms of a sentence are P and Q its universe is $/P/ \cap /Q/$. This intersection (the universe of discourse) is either empty or not. If empty then, by T.1, N(PQ), in which case the sentence is nonsense, category mistaken. If the sentence is category correct then its universe is nonempty. Now, given the law of category inclusion, when two categories do intersect, the intersection, is coextensive with one of them. Which one? The answer is provided by rule T.2.

(T.2) $U(PQ)U(QR)N(PR) \equiv (/P/ \subset /Q/)(/R/ \subset /Q/)(/Q/ \not\subset /P/)$
$(/Q/ \not\subset /R/)(/R/ \not\subset /P/)(/P/ \not\subset /R/)$

T.2 demands that, given two U-related terms with one higher than the other, the category with respect to the higher term properly includes the category with respect to the lower term. Consider again 'The white (thing) is a log.' The universe of discourse is $/\text{white}/ \cap /\text{log}/$ (the intersection of the categories determined by its two terms). Now, the sentence is category correct, U('white,' 'log'). And 'white' is higher than 'log,' /log/⊂/white/. Thus: /white/∩/log/=/log/. Our universe of discourse is the category of things with respect to 'log,' i.e., the set of logs or nonlogs. And, according to Sommers, it is this set of things which our sentence is about.

It seems that Sommers has two different theses about what a sentence is about. Both theses hold that what a sentence is about is not necessarily what satisfies its grammatical subject. But the first thesis (the natural subject thesis) claims that a sentence is about what its natural subject term is true of. Since the natural subject of 'This white (thing) is a log' is 'log,' this sentence is about logs. According to the second thesis (the universe of discourse thesis), a sentence is about its universe of discourse (which is empty for category mistakes and equivalent to the category with respect to the natural subject term otherwise). Thus 'This white (thing) is a log' is about the category of things with respect to 'log' (i.e., /log/). While the first theory says that our sentence is about logs (things of which 'log' is *truly* predicated), the second thesis says that it is about /log/ (things of which 'log' is *sensibly* predicated). Needless to say, the category with respect to 'log' is bigger than the set of logs.

3. The heavy emphasis on the subject-predicate syntax and the insistence on a distinction between the negation of a term and the contradictory of a sentence in Sommers' tree theory reveals an underlying logic at odds in important ways with the standard logic now entrenched in the schools. In 1967 Sommers began building an alternative logic (see chapters III and IV below). In the course of that work he offered his third account of what a sentence is about.

In chapter 10 of *The Logic of Natural Language* Sommers introduces his notion of *amplitude*. The amplitude of a term (in use in a given sentence) is determined by the "domain of application" of that term (in that sentence). A term may have different amplitudes in different sentences. And, while the amplitude of a term is not part of its meaning, multiple amplitudes in a given context have the same effect on inference as equivocity. A term is said to be used with "standard" amplitude when its domain of application is the actual world, existing things. Thus, in normal discourse situations, 'Secretariat is a horse' uses 'horse' with standard amplitude. That same term would normally be used with nonstandard amplitude in 'Pegasus is a horse.' Generally speaking, if 'P' is a term used with standard amplitude, and 'W_A' denotes the actual world, it is analytically true that every P is a member of W_A (= every P exists).

Given that in any normally used sentence a term has standard or nonstandard amplitude, what can we say of terms like 'exists,' or 'actual,' or 'is a member of the actual world,' and the like? Do they have any amplitude, and, if so, what kind? In chapter 10 Sommers claims that such terms have *no* amplitude (p. 216). But can this be right? There is no doubt, as many philosophers have emphasized, that terms like 'exists' are logically atypical. But to say that they have no amplitude would be to say that they are not used with respect to a domain of application. Thus, 'horse' in 'Secretariat is a horse' is used with respect to the domain of actual, existing horses, and so has standard amplitude; and it is used in 'Pegasus is a horse' with respect to the domain of mythological beasts, and so has nonstandard amplitude. But, if Sommers is correct here, in 'Secretariat exists' the predicate is not used with respect to any domain at all.

In chapter 15 Sommers seems to confuse the issue more by saying that terms like 'exists' have *unrestricted* amplitude (p. 335). And he clearly means that a term has unrestricted amplitude when it is used with respect to an unrestricted domain of application. In other words, in such cases, the domain of application for the term is not restricted to the actual world, or the world of Greek mythology, or any other conceivable world, but instead encompasses *all* such worlds. To say that tame tigers exist is not to characterize tame tigers (Kant's point) but to locate them in one of the worlds (viz. the actual one). Terms of unrestricted amplitude might be said to have *omniworld amplitude*. But notice now that such terms cannot be said to have, following Sommers of chapter 10, no amplitude.

Let us begin to straighten out these various ideas about amplitude in the following way (keeping in mind that we are free to use a given term, e.g., 'horse,' with different amplitudes in different sentences). Every term is used with respect to a domain of application. What that domain is on any occasion of the use of the term is a matter of shared pre-analytic knowledge. It is a piece of extralinguistic knowledge (or belief) which forms part of the background for that use of that term. Every term is used with either restricted or unrestricted amplitude. Its amplitude is restricted whenever its application is taken to be restricted to a particular domain. Otherwise it is unrestricted. I tell a small child that Pegasus does not exist because he or she wants to know the normal domain of applicability for 'Pegasus.' What I say, in effect, is that Pegasus is not in the actual world. In producing my sentence, in that context, I cannot assume as part of a shared background knowledge anything about the ontological location of Pegasus. The child just doesn't have this knowledge. On the other hand, you and I both know that Pegasus is a mythological beast so that, being aware of that, I can say that Pegasus is winged, and so forth, and while doing so restrict my domain of application to some particular world (viz. the world of Greek mythology).

Finally, every term used with restricted amplitude has either standard or nonstandard amplitude. Its amplitude is standard whenever it is used with the actual world as its domain of application. Otherwise, it has nonstandard amplitude.

Sommers has argued that it is a fallacy similar to equivocation to use a term in an argument without keeping its amplitude uniform. Thus the argument

Pegasus is a horse
Every horse is wingless
So Pegasus is wingless

is invalid since 'horse' has nonstandard amplitude in the minor premise but standard amplitude in the major premise. But the philosophically most interesting kinds of amplitudinal ambiguity involve arguments having a term used with restricted amplitude one time and unrestricted amplitude the next. Such arguments usually involve existential import or contain referentially opaque sentences. It would seem, for example, that, like 'exists,' modalized terms such as 'necessarily greater than seven' normally have unrestricted amplitude (see chapter IX below).

It is clear that, once properly formulated and understood, Sommers notion of amplitude is meant to supplant his previous notion of domain of discourse. If so, then one can conclude that the logical subject of a sentence, what it is about, is its amplitude. For example, 'Some men are in the Arctic' is about actual men, while 'Some men are in the Argos' is about mythical men. Finally, it should be noted that in his most recent work Sommers has replaced the term "amplitude" with "domain of discourse." We will add more on this notion in chapters IV and V below.

Chapter II

Endnotes

1. See Sommers (1970-71) and Englebretsen (1971a).
2. However, see Englebretsen (1972c), (1974c) and (1975c).

Chapter III

ON THE PHILOSOPHICAL INTERPRETATION OF LOGIC: ANOTHER ARISTOTELIAN DIALOGUE*

Dramatis Personae: Mr. Paleo, Mr. Neo, Mr. Anti, Mr. Conti, and the Aristotelian.

Scene: After making his views known to Mr. Paleo and Mr. Neo the Aristotelian had been driven from Academia.** It is now thirty years later. Mr. Paleo and Mr. Neo (the logic staff) and Mr. Anti and Mr. Conti (the philosophy staff) have heard rumors that the Aristotelian has returned. They seek him out and, finding him, are surprised by his unaged appearance.

Mr. Paleo: My dear Aristotelian, how good to see you again after all these years. I must say, you look well. The years have been especially kind to you.

Mr. Neo: Yes indeed. Hello, Aristotelian. You *do* look well.

Aristotelian: Thank you, my old friends. It is good to see you again after so many years. But your welcomes are too generous.

Mr. Neo: Not at all. You look as if you hadn't aged a day, sir. As you can see, we haven't fought Time nearly so successfully. Paleo has been in poor health for some time now. And I must admit I've been getting pretty run-down lately. But here now, let me introduce our new colleagues. This is Mr. Anti, of Oxford, and Mr. Conti, visiting us this term from Freiburg.

Aristotelian: It's a pleasure to meet you.

Mr. Anti: Mr. Aristotelian. I've heard a great deal of you.

Mr. Conti: And I too, sir. This is a pleasure indeed.

Mr. Anti: My friend, and frequent adversary, Mr. Neo here, tells me that he's heard rumblings about your return to logical studies. I must admit I have no stomach for his formal stuff. I hope you're not joining his ranks.

Aristotelian: Actually, Mr. Anti, my logical studies have never ceased. Unlike my old friend Paleo, I've never believed that logic was a finished subject. So I continue to work at it and hopefully to make progress. As for Mr. Neo's "formal stuff," no, I no longer embrace it with the generosity I once had. But I should add that I do agree with his notion that logic, of any sort, must be formal.

Mr. Anti: Well, I mean all that hen-scratching. It's fine for mathematics. But what we are about is philosophy. The problems faced by philosophers inevitably arise in the medium of ordinary language. I see no way in which Mr. Neo's investigation of his artificial, logically "corrected" languages can help. No, what's wanted is a closer attention to our ordinary modes of discourse.

Mr. Neo: You see, Aristotelian, what is happening. They want to give up the rigor of formal logic for the facile babble of the streets. You know, I still agree with your master. Logic is an organon of science. It formulates the rules governing correct inference and formal truth. Now we all know that a search for such a tool in ordinary language is futile. Indeed, as my master, Frege, said, logical work is a constant struggle against the logical defects of natural language. Thus, the logician's task of constructing a logically correct language.

Mr. Anti: Come now. What have all your fine systems done for us as philosophers? All these years of study to master the system only to worry over whether to put an exclamation mark after E. I tell you, it's logic which makes philosophy unattractive to the young.

Mr. Conti: Excuse me. But I must interrupt you here. It seems to me that you, Mr. Anti, are in no better position than Mr. Neo. You both sit in your towers picking at language...

Mr. Anti: *His* language is but a skeleton, a shadow of an artificial contrivance.

Mr. Conti: Skeleton-schmelleton. You both refuse to situate yourselves in Real Life. A philosophy must have a *Sitz in Leben*. Logic is dead--is Death.

Aristotelian: Well, I'm sorry to say that I can't agree with any of you.

Mr. Neo: You haven't changed.

Aristotelian: You, Mr. Anti, and you, Mr. Conti, would abandon formal logic. Yet formal logic is simply the code of reason. And surely reason is the philosopher's greatest treasure. It is logic, the laws of reason, which protect us from unreason, from philosophical nonsense, from inconsistency. You would abandon those laws. But when they're gone how will you preserve your reason?

Mr. Anti: But logic doesn't get us anywhere.

Aristotelian: But it's not a vehicle. Logic is a tool, used to protect our reason from invalidity and inconsistency. Your attitude reminds me of a line from Bolt's "A Man for All Seasons." Moore is defending the law (in this case the legislative, not the logical, kind) against those who would abandon it whenever it became inconvenient. He says: "And when the last law is down, and the Devil turned round on you--where would you hide, Roper, the laws all being flat?"

Mr. Conti: Well, it's not Reason which needs protection. It is Lived Life which is in danger. And Life must often go beyond Reason.

Mr. Anti: I don't know about that. But I must say, Mr. Aristotelian, it's not reason I'm against--just your logic. And especially Neo's logic.

Mr. Neo: Be careful. Don't put me with Aristotelian here. Those are nicely expressed sentiments you have, Aristotelian, but they are *sentiments*, I'm a logician. And one thing I'm sure of is that logic is best done without worries about reason. Reasoning is a matter for psychologists. My master, as you well know, has taught us all to keep logic and psychology apart--for the good of logic at least. No. I can't agree that logic is concerned with the laws of reason. Logic is the study of formal languages. That's it.

Aristotelian: When logic is taken from the hands of philosophers *that's* what becomes of it.

Mr. Anti: Indeed.

Mr. Paleo: I agree, my old friend, Aristotelian. You said a few moments ago that you no longer embrace Neo's logic the way you did thirty years ago. I'm glad to hear that. After all, we do quite well enough with Barbara, Celarent, Darii, and...and...my, to forget such things. My old memory is failing.

Mr. Neo: I'm sorry, Paleo. But, after all, it's no loss. The old logic is all easily incorporated into our system--and we don't need to learn any Latin.

Aristotelian: I'm afraid I've changed my mind on that, Neo. True, I once agreed with you that the old syllogistic could be fully incorporated into your logic. Though I believed it was still the most beautiful part of your system. But, now I don't think that is the case.

Mr. Neo: But how can you deny it? Every inference valid in the syllogistic can easily be shown valid in the standard predicate...

Aristotelian: Yes, yes. That I don't deny. But that just shows, to use your own terminology, that the old system is weakly equivalent to a part of your logic.

Mr. Neo: But surely, old friend, you don't want to claim that the old syllogistic is any more than this.

Aristotelian: To be quite honest, my work over the last several years has convinced me that the syllogistic system has the resources to compete on an equal footing with your logic as a system for reckoning all kinds of inferences.

Mr. Neo: What?!

Aristotelian: In fact, I claim further that it has advantages which strongly recommend it as a philosophical organon over yours.

Mr. Neo: He's mad!

Aristotelian: Perhaps someone should fetch help for poor Paleo there. I believe he's fainted.

....

Aristotelian: Let me explain to you why I make the claims I do. You, Neo, have been taught by your master, and in turn have taught so many others, that natural language is an unhappy medium for logical reckoning.

Mr. Neo: And so it is.

Aristotelian: You mathematical logicians, for whatever nonphilosophical reasons, have turned to mathematics, in particular, the syntax of algebra, for a model of logical form. What I mean is this. Since ancient times both grammarians and logicians believed that natural language sentences are structurally complex--consisting of a subject (noun phrase) and a predicate (verb phrase). Frege, however, with a distrust of the logical powers of natural language (healthy in a mathematician perhaps, but dangerous in a philosopher)...

Mr. Anti: Absolutely.

Aristotelian: ...turned to the functions and arguments of algebra as providing the best tools for an analysis of logical form.

Mr. Neo: Of course he did. After all, if you stick with subjects and predicates how on earth can you formulate relationals, or identities, or truth-functionals?

Aristotelian: Quite so, Neo. Nothing succeeds like success. And your logic *has* succeeded. By treating all sentences as functions on one or more arguments one can easily formalize relationals. By taking identity to be a relation between a thing and itself identities are easily added to the system. By allowing entire sentences to be arguments of truth-functions these too can be incorporated into the system. And this theory of logical syntax provides a solid foundation for the grand edifice of the standard predicate calculus.

Mr. Anti: A grand castle in the clouds.

Mr. Neo: Nonsense. It is, as you surely must admit, Aristotelian, the only universal logical calculus the realization of Leibniz's dream. How can you deny it?

Aristotelian: Well, I don't deny that it is an impressive, and useful, system.

Mr. Conti: Not so impressive. And Hegel's is far more useful.

Mr. Neo: We're talking about a *rigorous, formal* system here, Conti.

Aristotelian: Yes, indeed, Neo. You have achieved a very high degree of rigor. But are you so sure that the syllogistic system could not have achieved that same rigor?

Mr. Neo: Don't be silly. You know as well as I that the syllogistic achieved, at best, a kind of artificial rigor--Paleo's kind of rigor--nothing more. Besides, it's such a restricted system. It's hopeless as a candidate for a universal logic. Wedded as it is to the categorical form, it has no way to analyze relationals or identities, let alone truth-functionals.

Aristotelian: Yes, yes. You've been saying that for a hundred years now. But you haven't demonstrated that it is so. No. You merely claim that because syllogistic *did* not deal adequately with relationals and so on it *could* not. A nice *non sequitur*.

Mr. Neo: Well, Aristotelian, you did have twenty-four centuries.

Aristotelian: You're quite right, Neo. We have been slow. Leibniz almost succeeded in the Seventeenth Century, but...

Mr. Neo: Wait a minute! Surely you're not claiming Leibniz as one of your own. Everyone knows that Leibniz was the first mathematical logician. I remember Russell had only two portraits on his study wall: Frege's and Leibniz's.

Aristotelian: Yes, I know that's the common view held by the logical establishment today. But it's simply mistaken. Consider this. Leibniz never abandoned the idea that logical form was somehow based on the subject-predicate analysis. He never abandoned Aristotle (indeed, he chastised logicians who did), but urged only that the old syllogistic be strengthened and extended. Unlike Frege, he agreed with Aristotle that the logic of unanalyzed sentences was secondary to the logic of analyzed sentences, the logic of terms. And he never took logic to be the foundation of mathematics. No, Leibniz was a precursor of Frege only in that he too sought a universal logic, one embodied in a mathematical-like algorithm.

Mr. Neo: Well, perhaps. But you've admitted just why Leibniz failed where Frege succeeded--his Aristotelianism.

Aristotelian: Quite the contrary, Neo. His strength was his Aristotelianism. His failure was due mainly to his inability to find an adequate symbolic representation for sentences, and your old friend Russell told you how important an adequate notation can be.

Mr. Neo: So you think that syllogistic, provided with an adequate symbol symbolization, can compete with our standard calculus?

Aristotelian: Not just compete--but serve even better as a philosophical organon.

Mr. Neo: But how? It's too preposterous!

Aristotelian: Look, what's needed is (a) a sound theory of logical syntax which will incorporate all kinds of assertoric sentences, (b) a simple yet adequate system of symbolization, and (c) an algorithm on that system powerful enough to model all valid inferences.

Mr. Neo: The first order predicate calculus with identity.

Aristotelian: Quite so, Neo. But, what I have in mind is another system which achieves its advantages over yours by improving on (a).

Mr. Neo: Come now...

Aristotelian: According to your theory, every sentence is atomic or molecular (built up from atomic sentences by truth-functions and quantifiers). Atomic sentences always consist of exactly one function and an appropriate number of arguments. Functions and arguments are radically different syntactically. A function could never play the syntactic role of an argument, nor could an argument ever play the syntactic role of a function. Functions are syntactically complex. Arguments are not. Functions are incomplete (unsaturated); arguments are complete (saturated). Functions are predicational; arguments refer. An atomic sentence is achieved when a function is completed by an appropriate number of arguments. Higher functions then take atomic sentences as their arguments to form molecular sentences. A nice picture.

Mr. Neo: Indeed it is. After all, names and predicates--I mean, arguments and functions--are different from one another. We recognize this and reinforce it with a notational difference. Moreover, the theory accounts for the so-called "problem of propositional unity": What makes a sentence a unit rather than just a string of terms?

Aristotelian: Yes. Functions have holes; arguments are pegs fit to fill them. A unit is an object having no empty holes. Thus, sentences, as well as arguments, are complete, unitary, saturated

Mr. Neo: It's better than tying subjects and predicates together with copulae--such weak little things anyway.

Aristotelian: The picture is quite different from the one found in syllogistic. On that view there is no atomic/molecular distinction. Sentences are viewed as complexes of two syntactically dissimilar things--subjects and predicates. Both are themselves syntactically complex. A subject is a quantifier plus a term; a predicate is a qualifier (copulae, if you will) plus a term. That is why syllogistic recognizes no atomic sentences. The old syllogistic worked on the principle that subject-terms and predicate-terms are interchangeable. A term appearing in the predicate of one sentence could appear in the subject of another. On the other hand, subjects and predicates are *not* interchangeable--a point you often seem to miss.

Mr. Neo: All this is very interesting. But so what? Mathematical logic and syllogistic have, as we all know, very different theories of logical syntax. But the point is that a logic based on Frege's theory can analyze all kinds of sentences (and thus inferences) which the syllogist cannot.

Aristotelian: You are still as impatient as ever, Neo. In fact, it's that impatience which accounts for the premature ascendancy of your logic over syllogistic. Once Frege suggested a new logical syntax you discarded the old syllogistic with hardly a word.

Mr. Neo: Nothing succeeds like success.

Aristotelian: Look at your calculus. It's quite a bundle. You have, first of all, what you call "basic" logic--the sentential calculus. To this you add predicates (functions on names), pronouns (name variables), and quantifiers to get the first-order predicate calculus. Then you recognize that if identity is a relation it's a very special one, so you add identity (and all its accouterments) to get the first-order predicate calculus with identity--the standard system. Sentences like 'If it rains, it pours,' 'Every man is rational,' 'Tully is Cicero,' and 'Socrates is rational' are all given radically different logical analyses.

Mr. Neo: A small price for such a wide range of applicability.

Aristotelian: I said awhile ago that what I had in mind was a system which has inference powers comparable to your system, but which achieved an advantage over yours by embedding a superior theory of syntax. Let me explain.

Mr. Neo: I can't wait (though I don't expect satisfaction).

Aristotelian: Let's look at some sentences.

(1) Socrates is rational.
(2) Tully denounced Cateline.
(3) Every man is rational.
(4) Tully is Cicero.
(5) If it rains, it pours.

You would formalize these as follows:

(1.1) Rs
(2.1) Dtc
(3.1) $(x)(Mx \supset Rx)$
(4.1) $t=c$
(5.1) $r \supset p$

Mr. Neo: Yes, yes.

Aristotelian: Each, upon logical analysis, is seen to have a form different (usually quite different) from all the others.

And this in spite of the fact that some of them are, at least in English, apparently similar in form.

Mr. Neo: What's good for grammar isn't always good for logic. Never be fooled by appearance.

Aristotelian: Yes, well, the theory of logical syntax I have in mind, unlike Frege's, has two advantages. Frege's theory denigrates natural syntax (what you call "grammar"), and proudly offers what are unnatural forms for a variety of sentences. It does this by *translating* them from natural language into logical language. My theory preserves natural syntax and merely *transcribes* these sentences into a convenient notation which preserves natural syntax. Also, your theory must recognize several different logical forms for sentences. My theory treats all these in a syntactically simple and uniform manner. In short, my system is simpler and more natural than yours. This simplicity and naturalness is inherited by the calculus which is based upon it--a calculus which, in spite of its simple and natural formal basis, suffers no loss of inference powers.

Mr. Neo: I'll believe it when I see it.

Aristotelian: I can give you a very brief sketch of what I have in mind. The details have been worked out by my friend, Fred Sommers, and his pupils and colleagues. Let's begin with singular terms. You treat singular terms as being radically different from general terms. Indeed, since general terms are only fit for predicating, never for reference...

Mr. Neo: We must avoid Platonism, after all!

Aristotelian: ...they are formulated as first order functions (predicates). Since singular terms are fit for reference and never fit for predication, they are formulated as arguments. But how is the distinction made in the first place?

Mr. Neo: Well it's certainly not on the basis that singulars denote one object while generals denote many.

Aristotelian: So your friend Quine has said. And like him, you claim that the difference is a difference in "logical roles." But my question is: Why must singular terms and general terms be confined to different logical (syntactic) roles? I prefer to treat singulars and generals on a syntactic par. This means that either kind of term can occur either in a predicate or in a subject.

Mr. Neo: Nonsense!

Aristotelian: I can do this because, on my theory of logical syntax, a sentence can be viewed as a complex of a subject and a predicate, each of which consists of a term plus either a quantifier or a qualifier. Outside of subjects and predicates terms are all members of the same substitution class. You're right, I can't predicate singular terms. But I can't predicate general ones either. What I can predicate are predicates--qualified terms. And, again, subjects are never general terms. But neither are they ever singular terms. They are quantified terms.

Mr. Neo: Babble on.

Aristotelian: Any term, singular or general, can be either quantified or qualified (a syllogistic necessity). Now, admittedly, singular terms in subject position are virtually never found quantified in natural language. But this is easily explained. You will admit that each term has a denotation (at least, a purported one)?

Mr. Neo: Yes. But some *are* empty.

Aristotelian: Of course. Take the term 'philosopher.' Its denotation is Socrates, Plato, Aristotle, Aquinas, Bacon, Descartes, and so on. Now a term can be quantified either universally or particularly. Thus we can have, for example, 'every philosopher' and 'some philosopher.' Now, while terms *per se* denote, quantified terms *refer*. Universally quantified terms refer to their 'entire' denotation. Thus, 'every philosopher' refers to Socrates and Plato and

Aristotle and so on. Particularly quantified terms refer to an undetermined (though often determinable) part (though often all) of their denotation. Thus, 'some philosopher' refers to Socrates or Plato or Aristotle and so on.

Mr. Neo: This just sounds like poor Paleo's distributed/undistributed business.

Aristotelian: And so it is. Now notice that while on your theory reference is the role of singular terms--names, pronouns--on my theory reference is always the proper role of quantified terms. This means that singular terms, when used to refer, must be implicitly quantified.

Mr. Neo: Yes. Old Paleo was always making the silly claim that they were implicitly universal. He used to say that 'Socrates is rational' means 'every Socrates is rational.' Poor Paleo.

Aristotelian: Well, I think he was half right. Leibniz pointed out that we could just as well take singulars in referring positions to be particularly quantified. But he saw that it wouldn't make a difference since, for example, 'every Socrates' and 'some Socrates' make the very same reference--namely to Socrates. That's why natural language uses no explicit quantifiers for singular subjects. Quantifiers can be ignored there because they make no difference. Logically we can use whichever one we like. So it doesn't matter whether a term is singular or general. Every subject has the *logical* form: quantifier plus term.

Mr. Anti: A pretty cavalier way to treat the way we ordinarily refer.

Aristotelian: Treating subjects as always logically quantified terms permits both singular and general terms to play parts in reference. Treating identity not as a relation but as simply affirmation accomplishes two things: it permits both singular and general terms to appear in predicates, and it eliminates the need to force ambiguity on 'is.' We don't need two kinds of 'is' (one of predication and one of identity). The former will do for both.

Mr. Neo: You can't mean that! After all, look at 'Twain is Clemens.' For one thing the 'is' of identity is symmetrical--the 'is' of predication is not.

Aristotelian: But read 'Twain' as 'some Twain,' as Leibniz suggested we could. Then symmetry is preserved by the simple conversion of I propositions--and no 'is' of identity is needed. In fact, it can easily be shown that reflexivity and transitivity are also preserved for such sentences without ambiguating 'is.'

Mr. Neo: Hurumph!

Aristotelian: As you will.

Mr. Neo: What about relationals? Nobody can fit them to categoricals. And the same with the truth-functionals.

Aristotelian: Leibniz gave both a very good try.

Mr. Neo: And failed.

Aristotelian: And failed--but not by so much. Syllogists in the past, such as Leibniz, had tried to fit relationals to the categorical form by parsing relational predicates as complexes of monadic predicates. We know this will not work. What I propose in my theory is to recognize relational predicates as complex terms, and then to take all syntactically complex terms to have the logical form of an entire sentence. In other words, from the point of view of logical syntax, relational expressions consist of a subject and a predicate.

Mr. Neo: So, just like us, you propose to treat relationals as sentences with multiple subjects.

Aristotelian: Not at all. Every sentence consists, logically, of exactly one subject and one predicate. Relationals contain at least one complex term (usually the predicate-term), and complex terms, while not *being* sentences, have the logical form of sentences--a subject and a predicate. The subject of a complex term is in no way a subject of the sentence containing that term. Take the sentence 'Every boy loves some girl.' The subject is 'every boy,' the predicate is 'loves some girl.' And the predicate, being complex, can be taken as having the same syntax as a sentence--a subject, 'some girl,' and a predicate, 'loves.'

Mr. Neo: But how in the world could you keep the ordering straight?

Aristotelian: That can easily be taken care of in the notation system by indexing terms and by marking the degree of each relative term. Sommers has worked out an exceptionally simple, elegant symbolization for the whole system.

Mr. Neo: Even for truth-functionals?

Aristotelian: Yes, even those--though that's a name for them which is a bit prejudicial in your favor. Why not simply call them compound sentences?

Mr. Neo: Of course. But they *are* compounded by truth-functions. At any rate, I hope you're not going to try to tell me that compound sentences are really categorical. That would be too much.

Aristotelian: Well, in a way I do want to say that. But only in a way. You know, ever since the Stoic logicians there has been a debate about which logic is more basic--a logic of terms (i.e., a logic of simple categorical sentences), or a logic of compound sentences. Aristotle believed the latter was secondary to the former. The Stoics seem to have taken the contrary view. Leibniz clearly believed that compounds could be reduced to categoricals. Kant argued that neither could be reduced to the other. Many nineteenth-century logicians, included Boole and Frege, at least suggested a categorical reading for conditionals and conjunctions. Your logic takes the logic of compound sentences--your sentential, or truth-functional, logic--as basic to the predicate logic--the logic of terms. There is only one other possible thesis, and it was suggested by Pierce, who proposed to treat both categoricals and compounds as structurally isomorphic, sharing a common underlying syntax. I follow Pierce in this. What is required, then, is simply a system of notation which permits common logical forms for both categoricals and compounds.

Mr. Neo: But surely that's not possible.

Aristotelian: I haven't time to show you how the whole system works out notationally. But I can give you a brief example. In fact, I'll show you how I treat the five sentences we mentioned earlier, if you don't mind. That will give me a chance to illustrate my previous claim that the new system I have in mind is simpler and more natural,

Mr. Neo: Fine.

Aristotelian: The sentences, you recall, were

(1) Socrates is rational.
(2) Tully denounced Cateline.
(3) Every man is rational.
(4) Tully is Cicero.
(5) If it rains, it pours.

Now, on my theory, every sentence has the following structure: a subject and a predicate. Every subject has the structure: a quantifier and a term. Every predicate has the structure: a qualifier and a term. Every term is syntactically simple or complex. Complex terms have the syntactic structure of sentences. When the sentence is compound the subject-term and the predicate-term are complex. Not only do they have the structure of sentences--they are sentences. The vocabulary for talking about compounds is different, but these are mere verbal differences. u The structures are still the same. For example, 'if p then q' consists of a subject (antecedent) and a predicate (consequence). The subject consists of a quantifier ('if') and a term (sentence), and the predicate consists of a qualifier ('then') and a term (sentence).

Mr. Neo: Sounds funny.

Aristotelian: So now let me show you how we might begin to formulate our five sentences. We'll symbolize each term by an appropriate uppercase letter. Entire sentence structures are enclosed in square braces. Syntactical quantifiers are indicated by "qt"; qualifiers by "ql". Implicit formatives are placed in parentheses. So:

(1.2) [(qt) S ql R]
(2.2) [(qt) T (ql) [(qt) C (ql) D]]
(3.2) [qt M ql R]
(4.2) [(qt) T ql C]
(5.2) [qt [R] (ql) [P]]

As I said, Sommers has worked out a system of notation which is simple, elegant, natural, and perspicuous.

Mr. Neo: You said earlier that an acceptable system of logic must have an adequate theory of logical syntax, an adequate system of symbolization, *and* a sufficiently powerful calculus for inference analysis. Suppose you do have the first two--what about the last item. Surely it's the most important.

Aristotelian: Actually they are all equally important. After all, each is essential to the whole system. But the syntax must come first, then the notation, and finally the calculus. My algorithm is arithmetic by virtue of my symbolism; and it maps the syllogistic. I call it the "new syllogistic."

Mr. Neo: But why call it *syllogistic* at all? Syllogistic is just the old logic of categoricals.

Aristotelian: No. Syllogistic is a logic of sentences sharing the syntax of categoricals. Given Peirce's thesis, all assertoric sentences share this logical form. A logic or calculus is relative to a logical syntax. One could have a variety of syntactical forms, as you do, and thus complicate one's calculus (thus: sentential, predicate, identity), or one could have a simple, general syntactic form and then have a simple, general calculus. The syllogistic is a simple calculus based on the laws of conversion and the *dictum de omni et nullo*, which serves well for both categoricals and compounds, not because either is reducible to the other, but because they are structurally isomorphic.

Mr. Neo: I knew you'd start speaking Latin sooner or later.

Aristotelian: My friend, here is a sentiment you and I can both share:

Sicut se habet stultus ad
sapientem, sic se habet
grammaticus ignorans logicam
ad pertum in logica.

Postscript: As before, the conversation having become known to the authorities, the Aristotelian was again driven from Academia. Shortly thereafter Mr. Paleo died, Mr. Neo retired, Mr. Anti returned to Oxford, and Mr. Conti abandoned philosophy for politics. The work of the department is now in the capable hands of young Mr. Populo, well known for his papers on pollution, sex, and animal rights.

Chapter III

Endnotes

* A German version of this first appeared in *Zur Modernen Deutung der Aristotelischen Logik*, II, A. Menne and N. Offenberger, eds., Georg Olms Verlag: Hildsheim, Zurich, New York, 1985.

** See P. Banks, "On the Philosophical Interpretation of Logic: An Aristotelian Dialogue," *Dominican Studies*, III (1950); reprinted in *Logico-Philosophical Studies*, ed. A. Menne (Dordrecht, 1962).

Chapter IV

AN INTRODUCTION TO (A SOMMERS-LIKE) LOGIC

Introduction

A language consists of terms and formatives. Sentences in the language are combinations of terms and formatives, and are subject to the constraints of grammar. The grammar is a specification of all the conditions on sentence formation. Some of these "rules of grammar" are specific to that language, others are universal, applicable to any language. The terms of a language constitute its vocabulary, or lexicon. The lexicon may consist of a single homogeneous set or several distinct sets (each such set often being called a "part of speech"). Thus a lexicon might divide into nouns, verbs, pronouns, adjectives, adverbs, etc. The subdividing of a lexicon is language-specific. There is no universal constraint demanding any language's lexicon be subdivided.

The logician specifies the universal *logical* constraints on any language. The input for logic, then, is universal grammar. The logician has no interest, qua logician, in any specific language. Thus, he or she can ignore language-specific rules of grammar and language-specific lexical categories. Properly speaking, the logician attempts to account for all inferences conducted in the medium of any language. Some of those inferences depend wholly or in part on semantic features of various terms; others do not. *Formal* logic attempts to account for the latter kinds of inferences. A logic dealing with the former kinds is not "informal." (Informal logic is a special area of study with which we have no concern here.) We can think of a logic which takes into account inferences dependent upon semantic features as a "semantically interpreted," or material logic. The line between formal and material is not always easy to draw.

Traditional logicians normally divided their inquiries into three parts. Each part was meant to examine a distinct "act of the intellect," as the scholastics called it. These three acts were apprehension, composition (and division) and reasoning. Apprehension was the grasping or understanding of the ideas or concepts conveyed by the use of a term. Composition (and division) was the formation (subject to logical constraints) of sentences from the lexicon and list of formatives. Reasoning was the process of inferring one sentence from one or more other sentences. The idea here was that Aristotle had initiated this three-fold study by writing three separate works of the *Organon*, each of which dealt with one of these three intellectual acts. They were, respectively, *Categories, De Interpretatione* and *Prior Analytics*. In this chapter we will retain this old-fashioned outline for logic, dealing first with *semantics*, an examination of the semantic features of the lexicon and logical formatives; next with *syntax* in which we examine the ways in which lexical and formative elements are combined to form more complex expressions (particularly sentences); and finally with *syllogistic*, where we examine the logical constraints on inferences.

I. *Semantics*

We begin by making a distinction between terms as lexical items and terms as used in sentences. Terms when used in sentences are always *charged*. They are positive or negative. They have *polarity*.

English examples are 'tactful'/'tactless' and 'colorful'/'colorless.' The 'ful' and 'less' are overt signs of term polarity. In natural languages overt signs of negativity are ubiquitous. But such signs for positivity are very rare. As in arithmetic, positive polarity is almost always implicit. As used in a natural language sentence, a term with no explicit charge is always taken to be positive. English examples of such positive/negative pairs are 'red'/'nonred,' 'married'/'unmarried,' 'visible'/'invisible.' In formalization we will supply each term variable in a formula with an explicit sign of polarity (e.g., for 'red'/'nonred': +R/–R).

Terms, qua lexical items, are uncharged. While they are potentially charged, they are actually neither positive nor negative. Think of socks in a drawer. As drawer items, no sock is either a left or a right sock. Any such sock is potentially either. But, as used (worn), each sock is "charged"; it is either a left or a right sock. So, as items in the lexicon, terms are neither positive nor negative. We can think of terms in the lexicon as having "absolute" value, à la arithmetic again, and symbolize them accordingly. Thus, for the lexical entry 'red' we could write '/red/.' In general then, '/ϕ/' is the term 'ϕ' qua lexical item. Absolute terms, e.g., '/ϕ/,' have no polarity, are uncharged.

Each lexical item has both an intensional and an extensional semantic association. In the former case each absolute term is associated with a *feature*, which it *comprehends*. In the latter case each absolute term is associated with a *category* of individuals, which it *spans*. A feature is what the scholastics often referred to as a "determinable." A determinable such as *color* has as its "determinants" such things as *red, blue, yellow*, etc. We shall call these *properties*. Whatever has any of the properties *red, blue, yellow*, etc., has the feature *color*. Notice that whatever 'red' is true of has the property *red*. Whatever 'nonred' is true of has either the property *blue* or the property *yellow* or.... Let us say that 'red' and 'nonred' are "logically contrary" terms. Pairs of oppositely charged terms, e.g., +ϕ/–ϕ, are logically contrary. Terms like 'red' and 'blue' are merely contrary. A negatively charged term will always be equivalent to the disjunction of all the terms merely contrary to the corresponding positively charged term. Thus: 'nonred'='blue or yellow or...'; likewise: 'nonblue'='red or yellow or....' The properties *red* and *blue* are incompatible--no individual can simultaneously be both red and blue in color. The properties *red* and *nonred* are logically incompatible. Whatever has a given property or any property incompatible with it has a given feature. So, whatever is red or nonred is colored, has the feature *color*. We could simply identify the feature color, then, by '/red/'--or even '/blue/,' since it too determines the feature *color*. In the same way that '/red/' (or '/blue/' or '/yellow/,' etc.) is associated with the feature *color*, '/square/,' '/round/,' '/triangular/,' etc., are associated with, i.e., comprehend, the feature *shape*, and '/soft/,' '/rough/,' '/smooth/,' etc., comprehend *texture*. Any number of absolute terms, lexical items, might comprehend a given feature.

All the individuals which share a given feature are spanned by the terms comprehending that feature. Thus, all colored things (material objects, presumably) are spanned by such terms as '/red/' and '/blue/.' The set of all things spanned by a given term (or set of terms), i.e., sharing a common feature, is a category. For example, '/red/' and '/blue/' both determine the category of colored things, material objects. Terms which span the same things, such as '/red/' and '/blue/,' are "categorially synonymous." If we let /φ/ be the category of things spanned by '/φ/,' then 'φ' and 'ψ' are categorially synonymous if and only if /φ/=/ψ/.

An absolute term, an uncharged term in the lexicon, has these two semantic characteristics, then. It comprehends a feature and spans those individuals having that feature (such individuals constituting a category). Categories are determined by features, which are in turn determined by absolute terms. The study of categories is a part of ontology. (See chapter I above for more on ontology.)

Terms as used in sentences are never uncharged. They are always either positive or negative (though often in natural language sentences the polarity, especially when positive, is tacit). We saw that a term, qua lexical item, has both an intensional and an extensional characteristic. A term, qua sentence item, likewise has two such characteristics. Terms *signify properties*, and they *extend* over *individuals*. The set of individuals over which a term extends is its (proper) extension. If 'ϕ' is a term used in a sentence, we will say that it signifies the property ϕ-ness (or [ϕ]), and has in its extension whatever is ϕ, i.e., whatever has the property [ϕ]. Thus, just as the category whose members are spanned by a given absolute term is determined by the feature which that term comprehends, the extension of a given charged term is determined by the property which that term signifies.

The terms of a charged pair always signify properties which are incompatible. Such terms are logically contrary. The (implicitly positively charged) 'red' signifies the property of redness, [red], and extends over whatever is red, whatever has that property. The term 'nonred' signifies nonredness, [nonred] (or [$\overline{\text{red}}$]), and extends over whatever is nonred (i.e., whatever is blue or yellow or...). The positive term 'wise' signifies the property of wiseness, wisdom, [wise], and extends over whatever is wise.

When I say, in appropriate circumstances, that some horses when fully grown are shorter than me I am clearly talking about horses in the actual world. This is simply understood by me and my audience. When I say, in a literature class, for example, that a horse captured by Bellerophon was winged I am clearly talking about a horse in Greek mythology and not about any horse in the actual world. Let us say that each use of a sentence in order to make a statement is made, implicitly, relative to some specifiable *domain of discourse*. Thus, the domain of my first statement above was the actual world; the domain of my second statement was the world of Greek mythology. Domains are nonempty, compossible totalities of individuals. The actual world, possible worlds, worlds of fiction or imagination, and sets are all candidates as domains of discourse. As we will see, the truth or falsity of a statement is directly dependent upon the domain relative to which it is made. For example, whether 'Some horse has wings' is true or not depends upon the domain of discourse relative to which the sentence is used to make a statement. What that domain is will be determined usually by the context of the utterance.

The domain relative to which a statement is made has a semantic effect on the terms used in the sentence being used to make that statement. Its effect is to limit the extension of each used term. Suppose, relative to the domain of Greek mythology, I say 'Some horse has wings.' The term 'horse' has in its extension all horses--actual, fictitious, imaginary or whatever. But in my statement-making use of this sentence here my understood domain, the world of Greek mythology, limits the horses I happen to be talking about to just those in my domain. Let us say the denotation of a term used in a sentence is the intersection of its extension and the domain relative to which it is used. In this example the term 'horse' denotes all the horses in Greek mythology. Since every term used in a sentence has an extension, and since every statement-making use of a sentence is made relative to a specifiable domain, every term used in a statement-making sentence has a specifiable denotation.

Here are some examples of term denotations. The term 'men' in 'Some men are logicians' denotes all actual men when that sentence is used to make a statement relative to the actual world. The term 'wasp' in 'Some wasp wears a wig' denotes the wasp met by Alice when that sentence is used relative to Lewis Carroll's Looking-Glass world. The term 'prime' in 'No prime greater than 2 is even' denotes all prime numbers when used relative to the set of integers. The term 'guest' in 'No guests were invited to spend the night' denotes all the people in my house (except me) last New Year's Eve when used relative to the domain of my 1987 New Year's party. Notice that in the last example there are members of the domain which are not denoted by 'guest.' I was in the domain. So were two or three party-crashers, a nosey neighbor and, I think, a great dane. Yet none of us are denoted by 'guest' in this case. And this is so in spite of the fact that, presumably, the party-crashers, neighbor and myself (though not the dog) are all in the extension of 'guest.' The intersection of that extension with the domain consists of only my guests that night.

We have seen, then, that every term used in a statement-making sentence has a specifiable denotation. Most used terms denote several individuals; some denote just one; others denote none. The first are called "general" terms; the second are "singular" terms; and the last are often called "empty" or "vacuous" terms. Used relative to the actual world, 'men' is general, 'Socrates' is singular, and 'unicorn' is vacuous. Whether a term is general, singular or vacuous is a contingent matter, dependent upon the extension of the term and the domain relative to which it is used. Recognizing this, we will make much less of the three-fold distinction than is normally the case among logicians. In particular, our syntax will not distinguish between any of these. The distinctions here are purely semantic, not syntactic.

Every term used in a statement-making sentence has both a denotation and a signification. Moreover, as we will see in our discussion of syntax, every term so used is either quantified or qualified. Quantifiers and qualifiers (copulae) are formative elements. Quantifiers are either universal or particular; qualifiers are either affirmative or negative. English paradigmatic examples of each of these respectively are: 'every,' 'some,' 'is,' 'is not' (other examples are: 'all,' 'each,' 'one,' 'a(n),' 'are,' 'were,' 'do,' 'isn't,' 'aren't,' 'are not'). Quantified terms are logical *subjects*. Qualified terms are logical *predicates*.

Subjects *refer*. The reference of a subject is determined by its denotation and its quantity. Consider the logical subject 'every logician' as used in such statement-making sentences as 'Every logician is rational' and 'Max admires every logician.' Used relative to the actual world, 'logician' here denotes all actual logicians (Aristotle, Ockham, Leibniz, Boole, Frege, Russell, Quine, etc.). And the expression 'every logician' refers to all of them. (The scholastics said that a subject referring to its entire denotation was "distributed.") Now consider 'some logician' as used in 'Some logician is a mathematician' or 'Max wants to meet some logician.' Again, used relative to the actual world, 'logician' denotes all actual logicians. The logical subject 'some logician' refers to some (perhaps all) of them. (The scholastics said that a subject referring to an indeterminate, though perhaps determinable, part, though perhaps whole, of its denotation was "undistributed.") Every logical subject, quantified term, refers either to its entire denotation or to an indeterminate part of its denotation. Notice that in reference the signification of the term plays no direct role. We might think of quantification as "masking" a term's signification.

Predicates *characterize*. Predicates are positively or negatively qualified terms. They characterize the referents of the subjects to which they are syntactically attached as either having or lacking the properties they signify. Thus, a logical predicate 'is ϕ' characterizes the referent of the subject to which it is attached as having [ϕ]. The predicate 'isn't ϕ' characterizes the appropriate referent as lacking [ϕ]. The predicate 'is nonϕ' characterizes its appropriate referent as having [nonϕ]. The characterization due to a logical predicate is determined by its signification and its quality. Predicate terms have denotations (just as subject terms have signification). But the denotation of a qualified term is masked by its quality.

Although it has been the fashion among most logicians for some time now to reflect the semantic distinction between singular and general terms (and now and then empty terms) in the logical syntax, we have resisted the temptation to do so. Singular terms are on a syntactic par with general terms. This means that they can be logically quantified (play the logical role of subject) and they can be qualified (play the logical role of predicate). Singular terms differ from general terms only in that their users know, or take it, that they denote just one individual. Singular terms, like any term, denote individuals. When quantified, a singular term refers to the whole or a part of its denotation, depending on its quantity. But in either case the reference is the same (e.g., some Socrates=every Socrates). This is because a subject like 'some Socrates' refers to a part of the denotation of 'Socrates'--but Socrates is the *only* part of that denotation. Thus, 'some Socrates' refers to just what 'every Socrates' refers to--Socrates. That explains why even though singular subjects are always logically quantified they are never explicitly quantified in natural language. The logical quantity simply makes no difference. When formulating singular subjects we will be free to choose either logical quantity for singular subjects depending on our logical needs at the time.

Singular terms, like any term, signify properties. When qualified, a singular term characterizes the referent of the subject to which it is syntactically attached as having or lacking the property it signifies. For example, 'Socrates' signifies the property of being Socrates, [Socrates]. The predicate 'is Socrates' characterizes the referent of its subject as having [Socrates]. The predicate 'isn't Socrates' characterizes the referent of its subject as lacking [Socrates]. The predicate 'is nonSocrates' (= 'is someone other than Socrates') characterizes the referent of its subject as having [nonSocrates]. The recognition of singular predicates eliminates the need for a special logic of so-called "identity" statements. Statement-making uses of sentences such as 'Tully is Cicero' simply characterize Tully as being Cicero, as having [Cicero], just as 'Tully is Roman' characterizes Tully as being Roman, as having [Roman]. Copulae, such as 'is,' are always qualifiers. They need never be construed as special binary relational expressions. There is no 'is' of identity, only an 'is' of predication.

We can summarize our semantic theses thus far as follows. Qualified terms, logical predicates, characterize by virtue of their logical qualities and what their terms signify. Terms signify properties, which are determinants of features. Features are comprehended by absolute terms in the lexicon. Such lexical items span categories of individuals. An absolute term spans whatever has the feature it comprehends. Charged terms, as used in sentences, extend over whatever has the properties they signify. And they denote those individuals over which they extend which are in the domain relative to which they are used. Finally, a quantified term, a logical subject, refers to its entire denotation or an indeterminate part of its denotation, depending on its quantity.

In part II below we will see that terms are always either simple or complex. Complex terms are always the result of applying a logical formative to a pair of terms. Only simple terms need be listed as elements of the lexicon, but all terms, whether simple or complex, share the same semantic associations. Complex terms, just like simple terms, have extensions, denotations and significations. When quantified, they refer, and when qualified, they characterize. Statement-making sentences are, from a logical point of view, complex terms. As such, they both denote and signify. When considering sentences as terms we will call them *sentential terms*. A sentential term denotes the domain relative to which it is used. Consider, for example, the sentence 'Some horse is winged,' used to make a statement relative to the world of Greek mythology. In such a case the sentential term denotes the domain, the world of Greek mythology. Similarly, 'Every man is mortal,' used relative to the actual world, would denote that domain. So terms in general, when used, denote individuals. The kinds of individuals denoted by sentential terms are domains.

Terms in general, when used, signify properties of individuals. So, sentential terms signify properties of domains in particular, sentential terms signify *constitutive* properties of domains. A constitutive property is a property which a totality has by virtue of its having or lacking a member or kind of member. Consider a stew. If it has carrots, salt, potatoes and water, but no beef, we could say that it is carroty, salty, potatoey and watery and nonbeefy. The first four of these are positive constitutive properties. The fifth is a negative constitutive property. Salty is a constitutive property of the stew because salt is a constituent of the stew. Nonbeefy is a constitutive property of the stew because beef is not a constituent of the stew. In general, if a φ-thing is a constituent of a totality, T, then T is φ-ish, has the constitutive property [φ]; if no φ-thing is a constituent of T then T lacks φ-ishness, T is un-φ-ish, un[φ]; and if a nonφ-thing is a constituent of T then T is nonφ-ish, [nonφ], [$\overline{\phi}$]. Notice that being [nonφ] and being un[φ] are distinct. The actual world has the constitutive properties [horse] and un[unicorn], and both [red] and [$\overline{\text{red}}$], but is not un[red].

As we will see, statement-making sentences, like any term, come in positively/negatively charged pairs. Such pairs of nonsentential terms are logically contrary. Such pairs of sentential terms are logically *contradictory*. A sentence is true when used to make a statement just in case its domain has the constitutive property which the sentence signifies. Otherwise it is false. A positive sentence characterizes its domain positively; a negative sentence characterizes its domain negatively. Let [S] be the constitutive property signified by the statement-making use of the sentence 'S.' If 'S'/'not S' are a pair of logically contradictory sentences used to make statements relative to a domain, D, then the first positively characterizes D as having [S], while the second negatively characterizes D as lacking [S], as having un[S].

Our discussion of logical syntax will show that all statement-making sentences can be construed as having one of four general logical forms: particular subject attached to a positive predicate, particular subject attached to a negative predicate, or the negations of these two. Examples of these four forms are

I Some S is P
O Some S isn't P
E Not: some S is P
A Not: some S isn't P

The constitutive properties signified in each case are

I [SP]
O [S$\overline{\text{P}}$]
E un[SP]
A un[S$\overline{\text{P}}$]

For example, an I sentence characterizes any domain relative to which it is used to make a statement as being SP-ish, as having as a constituent an S which is P. The domain is characterized by the presence in it of SP-things. An A sentence characterizes a domain relative to which it is used to make a statement as failing to be S$\overline{\text{P}}$-ish, as lacking [S$\overline{\text{P}}$], as not having as a constituent as S which is nonP. The domain is characterized by the absence from it of S$\overline{\text{P}}$-things.

Suppose, relative to the actual world, I use the following to make statements.

1. Some horse is winged
2. No horse is winged (= Not: some horse is winged)
3. Every horse is four-legged (= Not: some horse isn't four-legged)
4. Some logicians are rational
5. Quine is rational

The first characterizes the actual world as having a winged horse, and is false since the actual world has no such constituent. Had my domain been the world of Greek mythology it would have been true. The second characterizes the actual world as not having any winged horses, and is true since no constituent of this world is a winged horse. The third sentence, like the second, also negatively characterizes the actual world. In this case it characterizes the actual world as not having any horses that fail to be four-legged, and is false if there happens to be, say, a three legged horse in the world. The fourth characterizes the actual world as having rational logicians among its constituents, and is, presumably, true. As is the fifth sentence, since Quine is rational in the actual world. Letting α stand for the actual world, and using appropriate abbreviations, we can display the truth conditions, then, of our five examples like this.

1. α is [HW]
2. α is un[HW]
3. α is un[H$\bar{F}$]
4. α is [LR]
5. α is [QR]

To summarize what has been said now about sentential terms: such terms, like any used term, both denote and signify. They denote the domain relative to which they are being used to make a statement, and they signify a constitutive property. The constitutive property signified by a sentential term is determined by the terms used in the sentence. The truth condition for a sentence used to make a statement is that its domain has the positive or negative constitutive property which it signifies.

Given the notion of truth outlined above, one can see why vacuous terms are of little concern to us. The extension of 'unicorn' and the actual world have no common constituents. The intersection of that extension and that domain is empty. So any statement-making use of a sentence relative to the actual world using the term 'unicorn' is simply false. To see this let '...U...' be a sentence using 'unicorn.' Used to make a statement relative to the actual world, $\mathfrak{a}$, this sentence has as its truth condition: $\mathfrak{a}$ is [...U...], i.e., $\mathfrak{a}$ has a such-and-such unicorn as a constituent. But $\mathfrak{a}$ has no unicorn of any kind as a constituent. The sentence, as used to make a statement in this case, is simply false, its truth condition fails to hold. Generally, a statement-making sentence is vacuous if and only if it uses a term which is vacuous. A term used in a statement-making sentence is vacuous if and only if its denotation is empty. Denotation is domain-relative. Vacuous sentences are vacuous relative to a domain. Vacuous sentences, as used to make statements, are false.

Thus far we have considered the semantics of used terms in general, and have paid special attention to singular terms, sentential terms and vacuous terms. We will examine the semantics of one more kind of term before turning to the topic of logical syntax. Relational terms are terms which are syntactically associated with more than one other term in a sentence. We will see just how this is done in the next section. For now we are interested in the semantic associations of relational terms. Consider the inference

Every lover is happy
Some boy loves some girl
So, some boy is happy

For this inference to be valid, which it surely is, there must be an intimate logical connection between the two terms 'lover' and 'loves.' They are, *prima facie*, different terms. The first is just a simple general term. The second is a relative term, relating 'some boy' and 'some girl.' Nonetheless, we will say that these terms are, from a logical point of view, semantically tied in such a way that in the inference above they cancel out one another (in the way that middle terms cancel out one another, as we will see, in a valid syllogism).

How are 'lover' and 'loves' semantically tied? Suppose Al loves Betty. Here are some of the sentences we can use to make true statements. 'Al loves,' 'Al is a lover,' 'Al is a lover of Betty,' 'Betty is loved by Al,' 'Betty is loved,' 'Betty is beloved,' 'Al is one who loves,' 'Betty is one who is loved,' etc. We seem to have several different versions of a common relational expression here: 'loves,' 'lover,' 'lover of,' 'loved by,' 'loved,' 'beloved,' etc. Yet the same relation, we want to say, is involved throughout. We cannot deny that these are different terms. The terms 'lover of' and 'loved by' are clearly distinct, and are involved in inferences differently. We will have to exhibit these differences somehow in our syntax. But all these terms signify the same relational property. Let [love] be the property signified by each of the terms belonging to the family of terms including 'loves,' 'beloved,' 'lover,' 'loved by,' etc. An individual enjoys (or suffers) the property [love] by virtue of its standing in the love-relation, its being a lover or a beloved. And just as all the terms in this family have a common signification, they all have a common extension (and thus denotation for any given domain), viz. whatever either loves or is loved. Generally, then, a relative term and all of its converse forms all signify the same property (called a *relation*) and have the same extension. For example, 'gives to,' 'gives,' 'is given,' 'is given by,' 'is given by...to,' etc., all signify [give] and extend over whatever stands in that relation.

We have looked now at terms as lexical items and then as used in sentences. We have seen that while a term has certain semantic associations qua term, these associations are modified in various ways determined by such factors as the charge on the term and its role as a logical subject or predicate (i.e., by quantity or quality). These are matters of syntax. In the next section we will see how to construct the rules of logical syntax.

II. *Syntax: Elementary Expressions*

Aristotle began chapter two of *Categories* by distinguishing between expressions "involving combination" and expressions "without combination." And in the first chapter of *De Interpretatione* he tells us that propositions (sentences used to make statements) involve "combination and separation." Sentences are among the expressions which involve combination. Such expressions are complex in a way in which a single word, for example, is not. But, of course, not all complexity of expression is of logical import. In prying the logical form of a sentence from it the logician wants to formulate in such a way that only logical distinctions are revealed. The logician needs to decide just what and how much should be formulated. Some distinctions are merely grammatical (some of these are local, others are universal). The ones the logician is interested in are those which will make a difference to the inference relations in which the sentence stands to other sentences. In some logical contexts the internal formal structure of a complex expression, such as a sentence, will not make a difference. It can be safely ignored. In other logical contexts the full logical form must be revealed.

The formal logician reveals or exhibits the logical form of an expression by using variables, in the way common to mathematics. This is both convenient and necessary. Many different expressions share a common logical form. Symbolizing by use of variables allows one to ignore the various differences among such expressions while concentrating on just what is common to them. Moreover, in the absence of a perspicuous system of symbolization, an algorithm, a mechanical procedure applied to symbolizations to reveal logical relations, would be virtually impossible.

The logical form of natural language expressions can be revealed in one of two ways. The first way, the way favored by most logicians today, is to construct an artificial language, with a vocabulary of one or more types of symbolic variables and a few symbolized formatives. The "grammar" rules of this language, i.e., the rules that determine which combinations of symbols are expressions of the artificial language and which are not, exhaust the formation constraints on expressions in that language. In other words, all the grammatical rules are logical syntax rules. Such a language is logically explicit. Logical forms need not be revealed. Surface structure *is* logical form. And, although such a language is never actually used in discourse, it is extremely useful in that expressions of a natural language can be *translated* into it, thus exhibiting their heretofore hidden logical forms. The second way, the one we favor, is to simply *transcribe* natural language expressions into a notation designed to reveal and preserve the logical features of that expression. According to this method, the logical form of a natural language expression can be revealed without first translating it into an artificially constructed language. Those who transcribe rather than translate see natural languages as themselves logical languages. There is a "logic of natural language." In giving a logical transcription of a natural language expression we ignore those grammatical features which in no way determine the expression's logical form. (For, again, the logician is interested in formal inference relations, which are fully determined by the logical forms of expressions entering into those relations.)

The basic kinds of expressions entering into inference relations are sentences used to make statements (or truth-claims). We will use "statement" as short for "statement-making sentence." Statements are complex expressions (things "involving combination"). They have parts. The basic elements of a statement are terms, more particularly, charged items from the lexicon. Indeed, we will take any expression, other than formatives (having any logical role in inference) to be a *term*. There are two kinds of terms: *simple* and *complex*. Examples of simple terms in English are general nouns such as 'man,' 'dog,' 'philosopher'; adjectives such as 'rich,' 'wise,' 'red,' 'happy'; mass nouns such as 'water,' 'wool,' 'gold'; abstract names such as 'justice,' 'wisdom,' 'beauty'; intransitive verbs such as 'walk,' 'live,' 'run'; transitive verbs (relative terms) such as 'loves,' 'gives,' 'believes'; proper names such as 'Socrates,' 'Al,' 'Betty,' 'John Smith,' 'Mars'; and personal pronouns such as 'it,' 'he,' 'they.' With the exception of pronouns, uncharged simple terms constitute the lexicon of a language. Simple terms will, despite the grammatical categories mentioned above, be given a uniform logical treatment. Consequently, all simple terms will be transcribed uniformly. Specifically, we will transcribe a simple term by a single letter (usually an uppercase Roman letter, often the initial letter of the natural language term).

As we have already seen, terms used in statement-making sentences are always logically charged--positively or negatively. Negative terms are almost always so marked in a natural language like English (e.g., 'tact*less*,' '*un*happy,' '*non*commissioned,' etc.). Positive terms are almost never explicitly so marked (though there are some exceptions in English, e.g., tact*ful*,' 'hope*ful*,' 'ful*some*,' etc.). Again, a positive/negative term pair will be said to be logically contrary. Positive terms will always be transcribed with a prefixed plus sign; negative terms will always be transcribed with a prefixed minus sign. Thus, 'tactful' will be transcribed as '+T'; 'tactless' as '–T.'

Before turning to complex terms, it is important to note a prominent difference between our treatment of simple terms and the treatment generally found among most contemporary logicians. The distinction between singular terms (such as proper names) and general terms (such as common nouns) is, on our theory, a purely semantic one. It is in no way reflected by a syntactic distinction. Singulars and generals are given the same syntactic treatment here. Suffice it to say, the standard logic of today lays a heavy syntactic burden on this distinction. The difference between singular terms and general terms is all-important. They are given distinct logical treatments and, of course, are translated by quite different kinds of symbols.

The positive and negative signs on terms are not themselves terms. They represent formative elements in the natural language. Formatives, then, are of two kinds. Unary formatives apply to a single term. The signs of polarity indicating whether a term Is positive or negative are unary formatives. Binary formatives apply to pairs of terms, resulting in a more complex term. There are two kinds of complex terms: *predications* and *compound terms*.

Elementary statements are predications They consist, from a logical point of view, of a pair of terms (simple or complex) joined by a binary formative expression Such statements are commutable (i.e., the formative is a symmetric one). Let '+A' and '+B' be two terms, and let 'i' be our binary formative, then '+A i +B' will be the form of an elementary statement. In general, the result of writing an 'i' between two charged term transcriptions will be an elementary statement form. We can read the 'i' here as 'is true of some.' Thus '+A i +B' could be read as 'A is true of some B,' and '–P i +Q' could be read as 'nonP is true of some Q.' Any natural language sentence that can be paraphrased by an English sentence of the form 'X is true of some Y' will be considered an elementary statement. The following are English examples of such sentences.

1. Some boys are dirty
2. A baby is crying
3. There are politicians who are dishonest
4. Some senator is lying
5. Some nonlogicians are mathematical

We could paraphrase these as:

1.1 Dirty is true of some boys
2.1 Crying is true of some baby
3.1 Dishonest is true of some politician
4.1 Lying is true of some senator
5.1 Mathematical is true of some nonlogicians

and transcribe them as:

1.2 +D i +B
2.2 +C i +B
3.2 –H i +P
4.2 +L i +S
5.2 +M i –L

These predications are all elementary statements. They are sentential terms. Not all predications are sentential. *Relationals* are predications. A relational is a complex term consisting of a charged term combined with a charged term, which itself is either a relational or a simple relative term (in English, a transitive verb, preposition, an adjective plus preposition, or verb plus preposition). The combination here is again achieved by our binary formative, the i-operator. In the case of relationals we will write the relative term first and read 'i' simply as 'some.' Any natural language expression which can be paraphrased by an English expression of the form 'R some X,' where 'R' is a charged relative term and 'X' is a charged term, will be considered an elementary relational. The following are English examples.

6. loves some girl
7. on a horse
8. gave a party
9. unhappy with some employees
10. looks at a nonphilospher

These can be paraphrased and then transcribed as:

6.1	+L	i	+G
7.1	+0	i	+H
8.1	+G	i	+P
9.1	–H	i	+E
10.1	+L	i	–P

Not all relationals have just one relata. Polyadic relations are common and expressions for them are ubiquitous in natural language usage. Our syntactic theory accounts for such relationals by admitting relational predications to the position of a relative term. Consider, for example, the English expression 'gave a toy to a boy.' Here the relata term 'gave to...' appears to have two relata. Yet, as a predication, any relational must be construed as a pair of terms connected by the i-operator. We can achieve the desired results by pairing the relative term with one relatum and then letting the resulting complex relational term serve as itself the relative term of a second relational containing the remaining relatum. Thus:

11. gave a toy to a boy

is first paraphrased as

11.1 gave to some boy some toy

The expression 'gave to some boy' is then transcribed, à la 6-10, as '+G i +B.' This relational then is taken as a relative term, to be combined by the i-operator with 'some toy.' Using parentheses to mark relationals which are parts of other expressions, we get the following full transcription

11.2 +(+G i +B) i +T

Thus far we have looked at one kind of complex term, the predication. And we have discerned between sentential and relational predications. It is important now to remember that sententials and relationals are *terms*. This means they can occur wherever a term can occur syntactically. A term occurring in a sentential term might itself be a relational, for example, or vice versa. Predications containing one or more predications as terms are no longer elementary. Here are some English examples, along with their appropriate transcriptions (we use brackets to indicate sententials which are parts of other expressions).

12. believes some gods are immortal
+B i + [–M i +G]

13. Some boy loves some girl
+(+L i +G) i +B

14. Some senator gave a party for a dictator
+(+(+G i +D) i +P) i +S

We have allowed our i-operator to do the necessary formative work for both kinds of predications. It can do this because it has certain formal features (viz. symmetry) which are common to various kinds of natural language expressions (sentential terms and relationals). Perhaps surprisingly, such formal features are common to every logically complex elementary expression. This means we will find the i-operator doing its job of combining terms not only in predications but in compound terms as well.

Any pair of terms, whether they are simple, sentential, relational, or themselves compound, can be combined by the i-operator to form a compound term. An elementary compound term is a conjunction of two simple terms. In English the most common expression for conjunction is 'and.' A term like 'old and wise' is an elementary compound. It conjoins two simple terms. The important formal feature of conjunction is that it is symmetric. Thus, compounds are commutable ('old and wise' converts to 'wise and old'). We can therefore use our i-operator to represent the conjunction formative. When a compound is part of a larger expression we will put it in angular brackets. Here are some examples of sentences containing compounds, along with their transcriptions.

15. Some philosopher is old and wise
+<+0 i + W> i +P

16. A boy kissed and hugged a girl
+(+<+K i +H> i +G) i +B

17. Some man kissed a girl and won a bet
+<+(+K i +G) i +(+W i +B)> i +M

Often we use, in English at least, relative pronouns such as 'who,' 'which' and 'that' to form logically compound terms. For example, we might say: 'Some logicians are mathematicians who are philosophers,' or equivalently: 'Some logicians are mathematicians and philosophers.' Both would be transcribed as: '+<+M i +P> i +L.' We might also simply juxtapose two terms in order to form a compound. In English we usually do this with an adjective-noun pair. For example, we say: 'Some old men are wise,' which would transcribe as: '+W i <+O i +M>.' The conjoined terms of a compound need not both be of the same logical type. We might very well, for example, conjoin a simple term with a relational. Thus: 'Some men are happy and love a woman,' which we transcribe as: '+<+H i +(+L i +W)> i +M.'

Perhaps the most common kind of conjunction in natural language is between sentential terms. In English we usually conjoin sentential terms with 'and' (but we also use such expressions as 'but,' 'while,' 'however,' etc.). The conjunction 'Some senator is lying and some administrator is guilty' has the logical form: '+[+L i +S] i +[+G i + A].'

Let us summarize now the kinds of terms found among the statement-making sentences of natural language. We have seen that, syntactically, every term is either simple or complex. Moreover, every complex term is either a predication or a compound. Predications, whether sentential or relational, are always pairs of charged terms combined by the symmetric i-operator. Compound terms are also always pairs of terms so combined. In short, all of our complex terms thus far are pairs combined by the i-operator. Let us call all expressions which can be transcribed using just our term variables, signs of polarity, and the i-operator *elementary*. Obviously not all the expressions of interest to the logician are elementary. There are other formatives--but, as we will see, they are all defined in terms of elementary expressions.

The notation we have used in transcribing is in many ways unnatural. It certainly deviates often from the surface forms of our normal English expressions. We have reversed the usual order of terms in some cases (e.g., 'Some man is wise' is transcribed as '+W i +M'), and we have used a large number of signs of polarity which are often absent in the natural language expressions (e.g., 'Some men are happy and love a woman' is transcribed as '+<+H i +(+L i +W)> i +M'). More importantly, there are many natural language expression forms which we can only transcribe properly using defined formatives. The rectification of these shortcomings will lead us to a full account of the kinds of expressions of import to logical inference and a more perspicuous notational system.

III. *Syntax: The Full System*

Up to now we have treated all syntactically complex expressions as the result of combining pairs of less complex expressions by use of the symmetric i-operator. This formative can be used to transcribe such English expressions as 'is true of some,' 'some,' and 'and.' The inspiration to treat sentential predication, in particular, in this way is Aristotelian. Aristotle tended throughout the Organon to paraphrase natural language statements as pairs of terms connected by a single formative. He would read sentences like 'Some man is wise' and 'Every logician is rational' as 'Wise belongs to some man' and 'Rational belongs to every logician.' Here the formatives 'belongs to some/every' connect pairs of terms to form sentences. It is theoretically interesting and important that we can do this. But, for the purposes of logical reckoning, it turns out best to unpack the syntactical information found in a single term-pair connective. This is just what traditional logicians and grammarians did. Instead of treating a sentence like 'Some man is wise' as the term-pair 'man' and 'wise' connected by a single formative, they treated it as two terms, each of which was already individually modified syntactically by a logical formative. What they did, and what we will do, is take a sentence like 'Some man is wise' to be a *subject* attached to (by mere juxtaposition) a *predicate.* A subject is a quantified term and a predicate is a qualified term. The single Aristotelian term-pair connective was unpacked into two different formatives: a quantifier attached to one term to form a subject and a qualifier attached to the other term to form a predicate. What this means initially for us is that our i-operator, when used to form predications, will be read as 'some...is....' Here 'some' is a quantifier and 'is' is a qualifier. This "split formative" view holds for compound expressions as well. The i-operator is again split into two formatives, one going with each term. This is most easily achieved with the conjunction reading of 'i' by the expression 'both...and....' This will force us of course to treat 'both' as a quantifier and 'and' as a qualifier. But there is no harm in this as long as we realize that we are talking only of *logical* quantifier, qualifiers, subjects, and predicates. We can think of ourselves as extending the nomenclature of predication to compounds by virtue of the fact that both share the same formal features (e.g., 'some...is...' and 'both...and...' are symmetric).

It is reasonable to symbolize the qualifiers 'is' and 'and' by a plus sign. We often do so naturally anyway. But how shall we symbolize our quantifiers 'some' and 'both'? Assuming that we want to use plus and minus signs for all formatives, the symmetry of the i-operator dictates that these quantifiers be transcribed as plus signs as well. Thus 'Some man is wise' is logically equivalent to 'Some wise thing is a man' since '+M+W' is algebraically equal to '+W+M' (note that the i-operator is replaced now by the split formative '+...+...').

Also, 'Some man is old and wise' is paraphrased as 'Some man is both old and wise' (by splitting the formative), which is logically equivalent to 'Some man is both wise and old' because '+M+<+O+W>' is algebraically equal to '+M+<+W+O>.'

In the transcriptions above we have already begun to introduce some simplifications. In particular, several plus signs have been suppressed (just as they often are in mathematics). Consider the sentence 'Some man is wise.' The terms 'man' and 'wise' are positively charged, and they are properly transcribed as '+M' and '+W' respectively, 'Some' is transcribed as a plus, and 'is' is transcribed as a plus. The entire predication is a complex term, having, as any term in use must, its own charge (in this case a positive one), transcribed by a plus. The full transcription would contain five plus signs: '+[++M++W].' Our simplified transcription is achieved by following the reasonable convention among mathematicians of suppressing all unnecessary plus signs. An expression standing on its own will always be considered to be a sentential predication, and, if unmarked, a positive one. Likewise, terms in general when unmarked will be taken to be positive. So, omitting the first, third and fifth plus signs in our example (i.e., the positive unary formatives), and the brackets (assuming the expression to be sentential), we arrive at our simple formula '+M+W.' Notice that we will not suppress minus signs, the signs of quantity or quality. For example, the contradictory of 'Some man is wise,' viz. 'No man is wise,' will be transcribed as '–[+M+W].'

Earlier we noted three ways in which our i-operator system of transcription was flawed. It often reversed the usual order of terms in a natural language expression (especially in sentential terms); it included a large number of plus signs, and it ignored many kinds of natural expressions which could not be easily transcribed using the i-operator alone. Our split-formative system solves the first of these. While the order of terms can make a difference in i-operator expressions because the terms are logically homogeneous, the order is flexible (as in natural language) for the new system, since subjects and predicates are always syntactically distinct. Thus we know that, given the formula '+M+W,' 'man' is the subject-term, not because it comes first, but because it is quantified. The second improvement is achieved now by our convention of suppressing unnecessary signs of positive term polarity. The third improvement remains.

So far our system allows us to transcribe any complex term whose formatives are paraphrasable as 'some...is...,' or 'both...and....' But any logical system worth its salt must at least be able to give a logical account of natural language expressions paraphrasable in terms of the following kinds of phrases: 'every...is...,' 'only...is...,' 'if...then...' and 'either...or....' As it turns out, each of these can be defined in terms of the formatives we already have in hand. Consider a sentential term such as 'Every man is wise.' We know that it is the contradictory of 'Some man is not wise.' This latter is directly transcribed as '+M–W.' Its contradictory, then, must be transcribable as '–[+M–W].' By "driving in" the external minus sign (as in algebra) we get '–M+W,' which we can then take to be the direct transcription of our original expression, 'Every man is wise.' Notice here that the qualifier 'is' is still transcribed by a plus sign, and the new quantifier, 'every' is transcribed as a minus sign. Consider next the relational term 'loves every girl.' Its logical contrary is something like 'doesn't fail to love some girl.' This latter is transcribed as '–(–L+G),' which, after distributing the external minus gives us '+L–G.' From now on those natural language expressions paraphrasable as 'every' (e.g., in English such phrases as 'each,' 'all,' 'every one of the,' etc.) will be transcribed directly as a minus.

In English we often form complex sentential terms by use of the phrase 'only if.' This is a single formative which is placed between pairs of sentential terms to form compound terms. But we also have in English a split-formative version of this operator, viz. 'if...then....' A sentence of the form 'If p then q,' where 'p' and 'q' represent any two sentential terms, is the logical contradictory of 'Both p and not q.' The contradictory of the latter is directly transcribed as '–[+[p]+–[q]].' By external minus distribution we can reduce this to '–[p]+[q],' indicating that we can transcribe 'if' as a minus and 'then' as a plus. In general, an expression of the form '–α+β' can be read as 'every α is β' or as 'if α then β,' depending upon whether the expression is seen as a predication or a compound.

Another logically important formative which results in compound terms is the one paraphrasable by the English 'or.' This phrase operates on pairs of terms (sentential or otherwise) to form compound terms. But, again, like 'only if,' we have in English its split-formative version--'either...or....' The negation (contrary or contradictory) of an expression of the form 'either α or β' is 'neither α nor β,' which is equivalent to 'not α and not β.' This latter is transcribed directly as +–α+–β,' so its negation will be '–[+–α+–β],' which reduces to '– –α– –β.' Here are some examples of English sentences and their transcriptions.

1. Some men are old or wise
   +M+<– –O– –W>
2. Every philosopher is stupid or some philosopher is wise
   – –[–P+S]– –[+P+W]
3. Some logician either reads or admires every book
   +L+(<– –R– –A>–B)
4. Whatever is either red or green is colored
   –<– –R– –G>+C

Finally, we consider here one other formative, the one paraphrasable by 'only.' In English 'only' is used along with a qualifier to form predications and with 'if' to form compounds. Examples are 'Only mathematicians are logicians' and 'Some logicians are wise only if all philosophers are logicians.' The first is equivalent to 'No nonmathematicians are logicians,' which, being the contradictory of 'Some nonmathematicians are logicians,' is transcribed as '–[+–M+L].' We can think of 'only,' then, as short here for 'not a non...' (or 'no non...'). Our second example above could be rephrased as 'Only if all philosophers are logicians are some logicians wise.' And this is equivalent to the contradictory of 'Both some philosophers are nonlogicians and some logicians are wise.' So it could be directly transcribed as '–[+[–[–P+L]]+[+L+W]].' In general, 'p only if q' has the form '–[+[–[q]]+[p]].' Whether as a formative in predications or in compounds, 'only' can be transcribed as 'Not some/both non/not...,' i.e., '–[+–...].'

Let us summarize our new notational system for transcription. We have two kinds of formatives, unary and binary. Unary formatives operate on each used term, simple or complex, charging it positively or negatively (indicated by a plus or minus sign). Binary formatives are always split into pairs of formatives, consisting in each case of a logical quantifier and a logical qualifier. Our elementary binary formative is transcribed as '+...+...' and read either as 'some...is...' or as 'both...and....' All other binary formatives are definable in terms of this elementary formative and our unary formatives. Every syntactically complex expression, then is viewed as a pair of terms (themselves simple or complex) connected by a binary formative pair. Also, from a syntactic point of view, we have seen that the distinction between predicational complexity and compositional complexity is an illusion. Both are the result of the combination of terms by means of a small and common arsenal of binary formatives. Indeed, any complex term can now be viewed simply as a concatenation of a *logical* subject and a *logical* predicate, i.e., a logically quantified term and a logically qualified term (where logical quantifiers and qualifiers encompass not only traditional quantifiers and qualifiers, but traditional compositional factors such as 'and,' 'if' and 'or' as well). The logical formation rules for any statement-making sentence can be easily expressed as follows. Logically:

(a) Every statement is a subject plus a predicate.
(b) Every subject is a quantifier plus a term.
(c) Every predicate is a qualifier plus a term.
(d) Every term is either simple or complex.
(e) Every complex term has the syntactic form of a statement.

IV. *Singular Terms*

We have already mentioned the fact that most contemporary logicians make much of the semantic distinction between singular and general terms. They reflect this distinction in their logical syntax. Traditional logicians did not do this. But they were, for the most part, far from clear about just how singular terms were to be treated logically. From our logical point of view, *all* terms are on a par syntactically. Any term, whether simple or complex, singular or general, relational, compound or sentential, can be quantified or qualified, and thus be fit for logical duty. In this section we will take a closer look at the syntax of singulars. We will see that they can be treated syntactically exactly like general terms, and that any extraordinary logical powers a singular might have are due to its semantics rather than its syntax.

The logical syntax of singulars is no different from the logical syntax of generals, Singular terms can be logically quantified and qualified, thus appearing both as subject terms and as predicate terms. Singular terms, as used in statements, come in logically contrary pairs, they are logically charged. Singular terms can be compounded with one another, or with other kinds of terms, to form complex terms.

We saw already, in section I above, that every term, including singulars, used in a statement has both a denotation and a signification. When universally quantified a term (now a subject term) refers to its entire denotation. When particularly quantified such a term refers to some undetermined part of its denotation. When the term being quantified is singular there is no difference between its entire denotation and any part thereof. There is but one part of the denotation of a singular term used in a statement. Consider the term 'Socrates' used in the statement 'Socrates is wise.' Its entire denotation (given the statement is made relative to the actual world) is just Socrates. And Socrates is the only part of that denotation. Thus, the quantification of 'Socrates' is, logically, arbitrary. 'Some Socrates' refers to Socrates. 'Every Socrates' refers to Socrates. So, since 'Socrates' is clearly the logical subject term of 'Socrates is wise,' and therefore must, on our theory of logical syntax, have a logical quantity (even if it happens to be tacit), the quantity makes no logical difference. We can transcribe singular subjects, then, as indifferently quantified. From now on we will use the asterisk, '*,' to indicate that the logical quantity is arbitrary. Our sentence 'Socrates is wise' would transcribe as '∗S+W.' Not all tacit quantity is arbitrary, however. Compare 'Babies cry,' 'Babies are crying' and 'Baby is crying.' The logical quantity of the first is universal ('–B+C'), the second is particular ('+B+C') and the third is arbitrary ('∗B+C').

The standard contemporary logic treats singular terms as paradigmatic subject expressions. Explicitly quantified general terms are then logically analyzed out of subject positions. We have treated explicitly quantified general terms as paradigmatic subject expressions admitting singulars into subject positions only with the admission price of tacit logical quantity. This second alternative is simpler than the first. And it requires far less deviation from the surface grammar of natural language sentences.

Consider the statement 'Tully is Cicero.' From the point of view of contemporary standard logic, this is an "atomic" sentence. As such it must consist of one predicate, or function, plus an appropriate number of singular term arguments. Well, the arguments are obvious--'Tully' and 'Cicero.' But what of the predicate? The stock answer is that in such cases as this 'is' must be the predicate. The 'is' here cannot, of course, be the same 'is' found in such sentences as 'Tully is Roman' or 'Tully is a senator.' In those cases the 'is' is the 'is' of *predication*. In the case of 'Tully is Cicero' the 'is' is the 'is' of *identity*. For the contemporary logician the 'is' of predication is logically inert, playing no role in the logical form of sentences in which it is used. By contrast, the 'is' of identity is logically all-important in sentences in which it is used. This 'is' is given its own symbolic translation and a special semantics, and special rules are devised governing inferences in which it is involved.

Suffice it to say, on our theory there is no predication/identity distinction. The copula, 'is,' is always a qualifier, forming logical predicates. In 'Tully is Cicero' and 'Tully is a senator' the expressions 'is Cicero' and 'is a senator' are both logical predicates. When used to make statements, both sentences are used to make a claim about what Tully is. In one case he is claimed to be a senator; in the other case he is claimed to be Cicero. In fact, these just are two of the things which he is, two of the properties he has. He is a senator, has the property of being a senator; and he is Cicero, has the property of being Cicero. On this theory, so-called identity statements are just predications whose predicate terms happen to be singular. We will transcribe a statement like 'Tully is Cicero' as '∗T+C,' a form quite like that for 'Tully is a senator, viz. '∗T+S.'

As terms on a syntactical par with any other term, singulars are logically charged, positively or negatively, when used in statements. Recall that two terms are logically contrary whenever they are polar opposites, alike except in having opposite charges. Two terms are merely contrary (contrary but not logically so) whenever they cannot both be affirmed simultaneously of the same subject. Thus 'red' and 'green' are nonlogical contraries. A general term like 'red' is logically contrary to 'nonred,' merely contrary to all other color terms (e.g., 'blue,' 'green,' 'yellow,' etc.), and is compatible with (not contrary in any way to) many other terms, i.e., 'square,' 'round,' 'happy,' 'old'). A singular term like 'Socrates' is logically contrary to 'nonSocrates,' merely contrary to all singulars with a different denotations (e.g., 'Aristotle,' 'Marx,' 'Mars'), and is compatible with many other terms (e.g., 'red,' 'green,' 'square,' 'old,' 'happy'). It is important to see why any two singulars with different denotations, say 'Socrates' and 'Aristotle' are (nonlogical) contraries. We have seen that two terms, 'P' and 'Q,' are contraries whenever nothing could simultaneously be both P and Q. Given that 'Socrates' and 'Aristotle' have different denotations, nothing could be both Socrates and Aristotle at the same time. Singulars such as 'Tully' and 'Cicero' are not contraries since they both denote the same individual, so that we can say of some one thing that is at once both Tully and Cicero.

The singulars 'Socrates' and 'Aristotle' are merely contrary. But 'Socrates' and 'nonSocrates' are logically contrary. This sounds nice--but what in the world is *nonSocrates*? Let us begin by admitting that the explicit negation of a singular term, such as 'nonSocrates,' is virtually never found in use in natural language. Nonetheless, we do ordinarily negate singular terms. To see how we do this, and what its logical effect is, we must note that *the negation of a singular term is not a singular term*. We say such things as: 'Socrates taught Plato, but someone else taught Aristotle,' 'Something other than Mars causes the primary perturbations of Mercury,' 'We saw every site but the Empire State Building,' 'The author of *Hamlet* was not Bacon.' The surface signs of negation here are 'else,' 'other than,' 'but,' and 'not,' and in each case they are intended to negate a singular. In the first case we intend by 'someone else' to make reference to some person (in the domain) who is nonSocrates (i.e., to some undetermined person not characterized by being Socrates). In the second case we refer to something in the domain which is nonMars. In the third we refer to every sight in the domain which happens not to be the Empire State Building. And in the last we characterize the author of *Hamlet* as being nonBacon, as being one who is not Bacon. The denotation of a singular is an individual in the domain of discourse. The denotation of its negation is every individual in the domain minus that individual. Suppose our domain is just the set of ancient Greek philosophers. While 'Socrates' denotes Socrates, 'nonSocrates' denotes all the other individuals in the domain (viz. Plato, Aristotle, Thales, Hereclitus, Parmenides, etc.). It follows that 'nonSocrates' is a general term, not a singular. And this holds for the negation of *any* singular.

There have been some who have argued that while we might very well form the negation of a singular term the result cannot itself be a genuine term, it cannot denote anything. The modality here is logical. The claim is that what would be denoted by a negated singular would be a logically impossible individual. Consider 'Socrates.' The argument has been that its negation would have to denote an individual, Nonsocrates, which has all the properties which Socrates lacks and lacks all those properties which he has. So, since Socrates is wise and Greek but not over six feet tall nor under four feet tall, not French, Russian, or foolish, Nonsocrates must, per impossible, be foolish, nonGreek, and both French and Russian, and both over six feet and under four feet tall. Of course, this argument against negating singular terms is empty once it is realized that the negation of a singular term is not a singular term.

The fact that singular terms are on a syntactic par with any other term means that they can be parts of compound terms. And again, their logical peculiarity here is due to their semantics, not their syntax. Terms are logically compounded by the (split) formatives 'both...and...' and 'either...or...' ('<+...+...>' and '<– –...– –...>'). Thus we form such compounds as 'old and wise,' 'either a member or an applicant,' and 'both men and women.' It is important to notice immediately that not every use of 'and' is aimed at forming a logical conjunction. Consider the sentence 'All the guests were men and women' (presumably no children were invited). We could not paraphrase this with the expression 'both men and women.' Let us say that whenever a predicate term of the form 'P and Q' can be paraphrased as 'both P and Q' it is a *genuine* conjunction, otherwise is is a *pseudo* conjunction. Pseudo conjunctions are, logically, disjunctions. They have the logical form 'either P or Q.' Thus 'All the guests were men and women' could not be paraphrased as 'All the guests were both men and women,' but rather as 'All the guests were either men or women.'

Every compound term whose constituent terms are singulars with different denotations will be a logical disjunction--so if it is a conjunction it is pseudo. Consider the sentence 'Socrates and Aristotle are wise.' Here 'Socrates and Aristotle' is a logical subject. It must therefore be implicitly quantified, But what is the quantity? Compare this sentence with 'Apples and oranges are fruits.' The quantity here too is implicit. Our choice is: 'All (things which are) apples and oranges are fruits' or 'Some (things which are) apples and oranges are fruits.' Clearly, we normally intend a universal reading here. But if so, then the compound term must be read as a pseudo conjunction. We could not paraphrase the sentence as 'All things which are both apples and oranges are fruits.' The proper paraphrase must be 'All things which are either apples or oranges are fruits,' which is transcribed as '<– –A– –O>+F.' Let us agree that just as nothing can be both an apple and an orange nothing can be both Socrates and

Aristotle. Every conjunction of 'Socrates' and 'Aristotle' will be pseudo. So what quantity is tacit in 'Socrates and Aristotle are wise'? First, let us paraphrase this so as to make the pseudo conjunction an explicit disjunction: 'Either Socrates or Aristotle is wise.' Once this is done it is obvious that the implicit quantity in our original sentence is universal: 'Every (individual which is) either Socrates or Aristotle is wise.' Suppose now our original had been an explicit disjunction in the first place, say 'Socrates or Aristotle taught Plato.' Then the implicit quantity would have been particular; 'Some (individual which is) either Socrates or Aristotle taught Plato.' Notice that the logical form of the compound singular in both cases is '<– –S– –A>.' Its logical quantity is universal when it is a pseudo conjunction, particular when it is an explicit disjunction, i.e.,

Socrates and Aristotle: –<– –S– –A>
Socrates or Aristotle: +<– –S– –A>

Some philosophers, under the influence of contemporary standard logic, have held a strange view of compound terms whose component terms are singulars. Orthodox logic has convinced them to take the subject of any subject-predicate sentence to be an expression making reference to a single individual. From this they have inferred that in sentences such as 'Socrates and Aristotle are wise' the subject must make reference to an individual having all the properties of both Socrates and Aristotle. Since there can be no such individual (for it would have to be at once both Macedonian and nonMacedonian, say), 'Socrates and Aristotle' cannot be a logical subject term. Indeed, the general conclusion is that compounds of singulars are, logically, not genuine terms. This conclusion, quite contrary to the evidence from natural language usage, is avoided by noticing that, in particular, conjunctions of singulars are always pseudo, and, generally, logical subjects need not be singular.

As a further explanation of these ideas we might consider more closely the semantics of singulars. Recall that each charged term as used in a statement, relative to a domain, has a denotation. Used relative to the actual world 'logician' denotes Aristotle, Leibniz, Frege,

Quine, etc. Suppose I want to denote just two logicians rather than all of them, say Russell and Whitehead. For example, I might make a statement like 'Russell and Whitehead wrote *Principia Mathematica*,' or 'The authors of *Principia Mathematica* were British,' or 'Either Russell or Whitehead taught Wittgenstein.' Now whenever a term is quantified it is used to make reference, and that reference is either distributed or undistributed, depending on whether all or only an undetermined part of the term's denotation is being referred to. Normally this is indicated by a quantifier. For example, 'every logician' makes reference to the entire denotation of 'logician,' while 'some logician' makes reference to an undetermined part of it. Suppose now that 'X' is a term whose denotation, relative to the actual world, is just Russell and Whitehead. It follows that 'every X' makes distributive reference to the denotation of 'X' (viz. Russell and Whitehead). And 'some X' makes reference to some undetermined part of the denotation of 'X' (viz. Russell or Whitehead). There are terms which have the denotation of 'X.' 'The authors of *Principia Mathematica*' is one such term. The statement 'The authors of *Principia Mathematica* were British' could be paraphrased as 'All (of) the authors of *Principia Mathematica* were British.' Here the subject refers distributively to the denotation of 'The authors of *Principia Mathematica*' (viz. Russell and Whitehead). The expressions 'Russell and Whitehead' and 'Russell or Whitehead' can both be used to make reference. The 'and' in the first case is a sign of distributive reference. The 'or' in the second is the sign of undistributive reference. We could make this explicit by paraphrasing these as 'All of (Russell or Whitehead)' and 'Some of (Russell or Whitehead).' This shows that in each case the logically quantified term is a disjunction. Pseudo conjunctions use 'and' not as a sign of conjunction (there is already a tacit sign of disjunction), but rather as an explicit sign of distributive reference, playing the same logical role as a universal quantifier.

If we had no term in English to denote logicians (in other words, terms like 'logician' would be excluded), we could still make reference, distributed or not, to logicians. Suppose we wish to state that all logicians are rational. In English-sans- ‘logician’ we could state ‘Aristotle and Leibniz and Frege and...and Kripke are rational.’ In order to state that some logicians are mathematicians we could use ‘Aristotle or Leibniz or Frege or...or Kripke are mathematicians.’ Here ‘and’ and ‘or’ are the signs of distributive and undistributive reference. We could paraphrase our statements as ‘All of (Aristotle or Leibniz or Frege or...or Kripke) are rational’ and ‘Some of (Aristotle or Leibniz or Frege or...or Kripke) are mathematicians.’ Notice that ‘Aristotle or Leibniz or Frege or...or Kripke’ has exactly the denotation of ‘logician.’ Indeed, we might think of *every* term as being, in theory at least, equivalent to a disjunctive term whose components are all singulars, each of which denotes one of the individuals denoted by that term. So if ‘P’ is a general term whose denotation is $A^1 \ldots A^n$ we could say: ‘P’=‘<– –A^1...– –A^n>.’ If ‘P’ is a singular term we could say: ‘P’=‘<– –P– –P>.’ All this is so because the denotation of any term is equivalent to the disjunction of the denotations of all the singular terms which denote each of the individuals in the term’s denotation.

Finally, notice that the explanation offered supports the notion that singular subjects are quantified arbitrarily. The expressions ‘Socrates and Socrates’ and ‘Socrates or Socrates’ are equivalent--both make reference just to Socrates. But the first is paraphrasable as’ ‘Every (Socrates or Socrates).’ And the second as ‘Some (Socrates or Socrates).’ And, since ‘Socrates or Socrates’ is equivalent to ‘Socrates,’ these become the equivalent pair ‘Every Socrates’ and ‘Some Socrates.’

V. *Syllogistic*

The theory of logical semantics and syntax sketched thus far is, of course, pointless unless it can satisfy the demand for an accurate account of logical reckoning. As logicians since Aristotle have recognized, logical reckoning is made most perspicuous by the use of a symbolic algorithm. Our symbolization can do just that once it is supplied with a few simple procedural rules.

Logical inferences are of two kinds, immediate and mediate (syllogistic). We begin with immediate inferences. Remember that the sort of inference of primary concern to the logician is *formal* inference. The inference of one statement from another is formal just when it depends upon nothing but the logical forms of the two sentences. The logical form of any sentence is determined by the theory of logical syntax which the logician applies to it. We have seen that on our theory the logical form of a sentence is revealed by transcribing it into our simple notation of term-variable letters, pluses and minuses (and various kinds of brackets for punctuations). Moreover, we found a surprisingly small variety of logical forms among the statement-making sentences of natural language. Each such statement consists logically of a quantified term and a qualified term. And any term may be either simple or complex. As well, complex terms themselves consist of a quantified term and a qualified term. Thus, our theory of logical syntax imposes a sometimes elusive but persistently uniform pattern on all statement-making sentences.

Given the notation of our full system, every sentence (used to make a statement, the only kind which enter into inferences for us) is positive or negative, has a logical subject which is particular or universal, has a (simple or complex) subject term which is positive or negative and a (simple or complex) predicate term which is positive or negative. The general form of any such sentence then, letting S and P be any terms (simple or complex), is:

+/–[+/– +/–S +/– +/–P]

Here the first plus/minus is the sign of affirmation/denial (the charge on a sentential term), the second is the sign of quantity, the third and fifth are signs of contrariety (the charge on nonsentential terms) and the fourth is the sign of quality. Signs of affirmation and positive signs of contrariety (in other words, positive term-charges) are often suppressed. For example, 'Every boy is lazy' is transcribed fully as '+[–+B++L].' But it can be transcribed more briefly, and usefully, as

'–B+L,' where the affirmation sign and the contrariety signs are suppressed.

We will classify every statement-making sentence as logically *P* or *U* depending upon the first two signs of its full transcription (i.e., the signs of affirmation/denial and quantity). A sentence whose first two signs are the same is P; a sentence whose first two signs differ is U. Notice that the contradictory of a particular/universal is always a universal/particular. Thus: the affirmation of a particular is P; the denial of a universal is P; the affirmation of a universal is U; the denial of a particular is U. In summary:

P	U
+[+...	+[–...
–[–...	–[+...

Since our system tends to reflect the syntax of natural language by suppressing most signs of affirmation, we can still determine whether a sentence in simplified transcription is P or U simply by algebraically distributing all external minus signs. In other words:

–[+/– +/–S +/– +/–P]=+[–/+ +/–S –/+ +/–P]

As an example, consider the sentence 'Not a creature was stirring.' Its full transcription is '–[++C++S],' which can be simplified as '–[+C+S].' This is a U sentence. By distributing the denial sign we get '+[–C–S],' which is simplified as '–C–S,' a U sentence which is explicitly universal. Simplification by suppression of affirmation signs and the distribution of denial signs reveals all P sentences as particular and all U sentences as universal.

Our first (of four) rule of immediate inference is (letting S_1 and S_2 be statements or their transcriptions):

> *Rule C*: There is a valid formal inference of S_2 from S_1 if (i) S_1 and S_2 are algebraically equal, and (ii) S_1 and S_2 are both P or S_1 and S_2 are both U.

Rule C encompasses all the traditional rules of conversion, obversion and contraposition.

There is another rule of immediate inference. Suppose we algebraically distribute to the right not only external minus signs (signs of denial) but all minus signs until our sentences are all logically affirmative and each simple term is in the range of just one sign. We will then call the sign on any term of such a formula its "distribution value." Every term of a sentence, then, has exactly one distribution value, determinable by algebraically distributing all minus signs as far to the right as possible. A term whose distribution value is negative is said to be *distributed*; a term whose distribution value is positive is said to be *undistributed*. Notice that two formulae are algebraically equal if and only if they share all the same terms (in any order) and any term distributed/undistributed in one is distributed/undistributed in the other. Our second rule of immediate inference is:

> *Rule S*: There is a valid formal inference of S_2 from S_1 if (i) S_1 and S_2 use all the same terms in the same order, (ii) there is a term universally quantified in S_1 and particularly quantified in S_2, and (iii) every other term has the same distribution value in both S_1 and S_2.

Rule S embodies the traditional rule of subalternation. Here are some simple examples of inferences justifiable by Rule S.

	S_1	S_2
1.	Every boy is lazy –B+L	Some boy is lazy +B+L
2.	Every boy kissed every girl –B+(K–G)	Some boy kissed every girl +B+(K–G)
3.	Every boy kissed every girl –B+(K–G)	Every boy kissed some girl –B+(K+G)
4.	Some boy kissed every girl +B+(K–G)	Some boy kissed some girl +B+(K+G)
5.	Every boy kissed some girl –B+(K+G)	Some boy kissed some girl +B+(K+G)

Notice that an inference of 'Some boy kissed some girl' from 'Every boy kissed every girl' is valid, but not immediate.

Strictly speaking, our third rule of immediate inference is not formal. Yet it is so useful, and inferences involving it are so common, that we include it here. Recall that singular subjects may be quantified either universally or particularly. More properly, such subjects are (implicitly) logically particular but from statements in which they are used we can immediately infer statements with the quantity universal. Thus we have

> *Rule I*: There is a valid material inference of S_2 from S_1 if (i) S_1 and S_2 use all the same terms in the same order, (ii) there is a singular subject in S_1 which is universally quantified in S_2 and (iii) every other term has the same distribution value in S_1 and S_2.

Rules I and S guarantee us the license to treat singular subjects as arbitrarily either particular or universal.

Our fourth rule of immediate inference involves relational terms. Relational terms are complex terms which are predicational but not sentential. As such they share the logical syntax of predications in that they consist of a quantified and a qualified term. The qualified term is called a "relative" term.

A relative term may itself be a complex relational term. Consider the sentence 'Some boy kissed some girl' ('+B+(K+G)'). The relational term here is 'kissed some girl' (in ordinary English, relational terms tend to retain the unsplit-formative construction--'some' here is our old i-operator). This relational consists of the relative term 'kissed' and the quantified term 'some girl.' Consider next 'Some executive gave every senator a bribe' ('+E+((G–S)+B)'). Here the relational term is 'gave every senator a bribe,' whose relative term, 'gave every senator,' is itself a relational term (consisting of the relative term 'gave' and the quantified term 'every senator'). From now on we will call quantified terms which are part of relational terms "logical objects." A logical object is just a logical subject occurring in a relational term. In 'Some executive gave every senator a bribe,' 'some executive' is the logical subject, and 'every senator' and 'a bribe' are logical objects. Relative terms can be viewed as tying together, in a specifiable order, subjects and objects.

Let us return now to our first sentence, 'Some boy kissed some girl.' From it we cannot infer 'Some girl kissed some boy' ('+G+(K+B)'). Yet Rule C would appear to justify such an inference. In order to avoid this we must recognize that relative terms like 'kissed' always tie their subjects and objects in a certain order. This order needs to be indicated in our transcription. An easy way of doing this is to give numerical subscripts to all subjects and objects, and then subscribe those numerals to the relative term in the order in which we wish the relative term to tie its subject and objects. For example, we can transcribe 'Some boy kissed some girl' as '$+B_1+(K_{12}+G_2)$,' indicating that the subject (1) kissed the object (2). 'Some executive gave every senator a bribe' could be transcribed as '$+E_1+((G_{123}-S_2)+B_3)$. Now we can see why 'Some girl kissed some boy' does not follow by Rule C from 'Some boy kissed some girl.' From '$+B_1+(K_{12}+G_2)$' we infer, by Rule C, '$+G_2+(K_{12}+B_1)$.' But this does not transcribe 'Some girl kissed some boy.' Instead, it transcribes 'Some girl is kissed by some boy.' 'Some girl kissed some boy' would be transcribed as '$+G_1+(K_{12}+B_2)$,' or perhaps '$+G_2+(K_{21}+B_1)$,' but neither of these follow by Rule C from '$+B_1+(K_{12}+G_2)$.' Notice that the inference from '$+B_1+(K_{12}+G_2)$' to '$+G_2+(K_{12}+B_1)$' is justified by Rule C. It is an instance of what linguists call "passive transformation." This same effect could be achieved simply by changing the order of subjects and objects while retaining the order of numerals subscribed to the relative term. Thus 'Some girl is kissed by some boy' is transcribed as '$+G_2+(K_{12}+B_1)$.' By contrast, 'Some girl kissed some boy' ('$+G_2+(K_{21}+B_1)$' or '$+G_1+(K_{12}+B_2)$') is not a passive transformation of 'Some boy kissed some girl' ('$+B_1+(K_{12}+G_2)$') because either (i) $K_{12} \neq K_{21}$ or (ii) $G_1 \neq G_2$ and $B_1 \neq B_2$. Here are some examples of passive transformations from 'Some executive gave every senator a bribe' ('$+E_1+((G_{123}-S_2)+B_3)$').

1 Some executive gave a bribe to every senator
$+E_1+((G_{123}+B_3)-S_2)$

2. Every senator was given a bribe by an executive
$-S_2+((G_{123}+B_3)+E_1)$

3. Every senator was given by an executive a bribe
$-S_2+((G_{123}+E_1)+B_3)$

4. A bribe was given to every senator by an executive
$+B_3+((G_{123}-S_2)+E_1)$

5. A bribe was given by an executive to every senator $+B_3+((G_{123}+E_1)-S_2)$

Many of the inferences we make involving relationals make use of these kinds of passive transformations justified by Rule C. But Rule C also justifies other kinds of inferences involving relationals. We might call these inferences "associative shifts." The inference from 'Some boy kissed some girl' to 'Something some boy kissed is a girl' is such a shift. It is achieved simply by shifting the parentheses enclosing the relational term. Transcribing, our inference looks like this:

$$\frac{S_1:\quad +B_1+(K_{12}+G_2)}{S_2:\quad +(+B_1+K_{12})+G_2}$$

As a second example, consider the inference of 'Someone whom a boy kissed is a girl who loved him' from 'Some boy kissed a girl who loved him.'

$$\frac{S_1:\quad +B_1+(K_{12}+<G_2+L_{21}>)}{S_2:\quad +(B_1+K_{12})+<G_2+L_{21}>}$$

Associative shifts and passive transformations are justifiable by Rule C. But a new rule is required to justify the following valid inferences.

(a) Helen is loved by Paris
So Helen is loved

(b) Some boy offered some girl a rose
So some boy offered a rose

These inferences are valid, but they are not immediate. Nonetheless, we can derive their conclusions by a series of immediate inferences. The final inference in each case will require our new rule. We can transcribe (a) as

$$\frac{+H_2+(L_{12}+P_1)}{+H_2+L_{12}}$$

Our first step is to infer from the initial formula '$+(H_2+L_{12})+P_1$' (i.e., 'What Helen is loved by is Paris') by Rule C (viz. associative shift). From this we infer our conclusion by a new rule.

We transcribe (b) as

$$\frac{+B_1+((O_{123}+G_2)+R_3)}{+B_1+(O_{123}+R_3)}$$

Here we first infer '$+B_1+((O_{123}+R_3)+G_2)$' ('Some boy offered a rose to a girl') by passive transformation (Rule C). Next we infer '$+(B_1+(O_{123}+R_3))+G_2$' ('What some boy offered a rose to is a girl') by associative shift (Rule C). Finally, we apply our new rule to get our conclusion.

Let us now formulate this new rule.

> *Rule R*: There is a valid immediate inference of S_2 from S_1 if (i) the logical subject of S_1 is a predicational term, i.e., relational or sentential, (ii) S_2 has the same form and uses the same terms as the logical subject of S_1, and (iii) S_1 is P.

This new rule is probably more familiar to us when we see it used on statements whose logical subjects are sentential terms. For example, from 'Some men are working but John is asleep' ('+[+M+W]+[+J+A]') we immediately infer 'Some men are working' ('+M+W').

Our four rules of immediate inference can be applied to statements one at a time to yield new inferred statements. An inference of S_2 from S_1 which requires more than one such application is not immediate. Yet not all nonimmediate inferences are the results of applying our rules of immediate inference more than once. Sometimes we can directly infer a statement from a *pair* of other statements. Such an inference is called "syllogistic." We have just one rule of syllogistic inference.

> *Rule DO*: There is a valid formal syllogistic inference of S_3 from S_1 and S_2 if and only if (i) S_3 algebraically equals the sum of S_1 and S_2, and (ii) either (a) S_1, S_2 and S_3 are all U, or (b) S_1 is P, S_2 is U, and S_3 is P, or (c) S_1 is U, S_2 is P, and S_3 is P.

Put simply, Rule DO just says that for a syllogistic inference to be formally valid the premises must add up to the conclusion and the number of P conclusions must equal the number of P premises.

Here are some examples of inferences satisfying Rule DO.

1. Some boy kissed every girl
Whoever kissed every girl is tired
So some boy is tired
$+B_1+(K_{12}-G_2)$
$-(K_{12}-G_2)+T$

$+B_1+T$

2. If anyone is lazy John is
Someone is lazy
So John is lazy
$-[+P+L]+[+J+L]$
$+P+L$

$+J+L$

3. Luna is a natural satellite of Earth
Every natural satellite of Earth is the moon
So Luna is the moon
$+L+N$
$-N+M$

$+L+M$

Every valid inference can be accounted for by one or more applications of Rules C, S, I, R and DO. Here is an example of a valid inference using all of our rules. Our premises are 'Every romantic admires any man who loves Helen' and 'Helen is a seductress.' From the second premise, by I, we infer 'Every Helen is a seductress,' from which, along with the first premise we infer, by DO, 'Every romantic admires any man who loves a seductress.' From this we infer, by C (passive transformation), 'Any man who loves a seductress is admired by every romantic.' Then from this we infer, by S, 'Some man who loves a seductress is admired by every romantic.' Finally, from that we infer, by R, our conclusion, 'Some man who loves a seductress is admired.' Symbolically:

1.	$-R_1+(S_{12}-<M_2+(L_{23}+H_3)>)$	premise
2.	$+H+S$	premise
3.	$-H+S$	2, 1
4.	$-R_1+(A_{12}-<M_2+(L_{23}+S_3)>)$	1, 3, DO
5.	$-<M_2+(L_{23}+S_3)>+(A_{12}-R_1)$	4, C

6. $+<M_2+(L_{23}+S_3)>+(A_{12}-R_1)$ 5, S

7. $+<M_2+(L_{23}+S_3)>+A_{12}$ 6, R

This brief sketch of the algorithm is hardly complete. Some details will be filled in later chapters. But, before concluding this chapter, we must point out that the system of rules for inference presented here is at least sufficient for analyzing all the kinds of inferences accounted for in the standard system of logic accepted by most logicians today. This means that all the usual inferences involving the standard propositional calculus, first-order predicate calculus, and identity theory can be analyzed by our simpler system. Indeed, we can say more. Some valid inferences beyond the limits of analysis offered by the standard system are easily dealt with by ours. We conclude, for now, with one simple illustration of this. The standard calculus has difficulty giving a perspicuous formal analysis of inferences involving singulars in relational sentences. The simple inference of 'Helen is loved by Paris' from 'Paris loves Helen' is hardly accounted for by a logical system which symbolizes both the premise and conclusion by 'L_{ph},' yet that is the best the standard system can do. We, of course, analyze this inference as a case of passive transformation, allowing logical subject and object to exchange places.

$$\frac{+P_1+(L_{12}+H_2)}{+H_2+(L_{12}+P_1)}$$

Now the standard logician can respond by claiming that his 'L' can be read as 'loves' in the premise but as 'is loved by' in the conclusion. And this, along with a reversal of the two argument terms ('p' and 'h') will give a conclusion formally distinct from the premiss. Thus:

$$\frac{L_{ph}}{L_{hp}}$$

But now the term 'L' is ambiguous, in which case the inference's validity can only be preserved by recourse to an extra-logical semantic relation between the two senses of 'L,' making the inference not formal at all. A system which takes passive transformation to be formal seems at once both simpler and more natural.

Other examples of inferences beyond the reach of the standard logic are

1. Socrates taught a teacher of Aristotle
   So one whom Socrates taught was a
   teacher of Aristotle

2. Brutes killed Caesar with a dagger
   So Caesar was killed

There is good common sense in favor of demanding that logic provide an analysis of such simple, everyday inferences. Our logic preserves this piece of common sense.

Chapter V

REMARKS ON THE SEMANTICS OF TERMS AND SENTENCES

Aristotle claimed that philosophy might be viewed as the art of making distinctions. With this in mind we present here some fundamental ideas and distinctions for a theory of the meanings of terms and sentences. The theory may not account for all of the meanings of such things, but it will have the advantage of being a "unified" semantic theory in that it gives a uniform account of at least part of the meaning of both terms and sentences. The unity comes from seeing sentences as syntactically complex terms. That part of meaning with which we are concerned consists of *signification* and *denotation*. The denotation of any expression will be seen to depend upon two things: its *extension* and the *domain* of discourse relevant to its use. The extension of an expression will, in turn, be seen as dependent upon its signification, which is itself related to the expression's comprehension. The semantic theory outlined here is intended to serve as a companion for the theory of logical syntax which has been much more fully and rigorously formulated by Sommers and his followers.[1]

1. Terms, as they are used in sentences, are either positive or negative--they are logically *charged.* Just as the positive charge in algebraic expressions is omitted (e.g., '+a x – b' is written simply as 'a x – b') the positive charge on terms in natural language sentences is usually suppressed. The negative charge is almost never suppressed. Thus we say, 'Some senators are unwise,' where the negative charge on 'wise' is explicit but the positive charge on 'senators' is tacit. The positive and negative versions of a term are *logically contrary* to one another.

Terms used in a sentence always *signify* properties. We could say that 'wise' signifies the property of wisdom while 'unwise' signifies the property of unwisdom. The property of unwisdom might be said to be just the property of foolishness. Likewise, the property of wisdom might be said to be just the property of unfoolishness. Now the term 'foolish' also signifies the property of foolishness or unwisdom. Thus we say that the terms 'foolish' and 'unwise' are co-significant in that they signify the same property. Likewise, 'wise' and 'unfoolish' are co-significant. Notice that given any positive/negative term there is (or we could construct) a corresponding negative/positive term signifying the same property--a co-significant term. We will say that the charge on terms is thereby reversible.[2]

If 'T' is a term of a sentence then [T] will be the property signified by that term. Thus: 'wise' signifies [wise], i.e., the property of wisdom; 'unwise' signifies [unwise]; 'foolish' signifies [foolish]; and so forth. Notice that 'foolish' and 'unwise' are cosignificant because [foolish] and [unwise] are identical properties.

Persons, plays and promises can be wise or foolish. Numbers, nails and nutrients can be neither. Let us say that that by virtue of which a thing can be said to have a property is a *feature* of that thing. There is, then, a feature which wise things as well as foolish things share, but which is not shared by things like numbers, which cannot be said to be either wise or foolish. When considering features we can ignore the charge on a term. For 'wise' and 'unwise' may signify different (indeed incompatible--because the terms are contrary) properties, but they specify in some way the same feature. Let us say that 'wise' and 'unwise' both *comprehend* the feature of being the sort of thing which can be said to be wise or unwise (foolish). Let /wise/, read "absolute 'wise,'" be the uncharged term, neither positive or negative. And let [/wise/] be the feature comprehended by /wise/, the feature of being the sort of thing which can be said to be wise or unwise. Then, since /wise/ and /foolish/ are the same feature, /wise/ and /foolish/ are comprehensively synonymous terms.

Absolute terms are not used in the sentences of ordinary discourse. Only charged terms are to be used. We can think of absolute terms as being the elements of our lexicon. Our lexicon then will consist of all those expressions which comprehend features and which, when charged and used in sentences, signify properties.

A term not only signifies a property but also *denotes* objects. What objects a term used in a sentence denotes depends upon two things: the *extension* of the term and the *domain*, or universe of discourse, relative to which the sentence is used. The domain of a sentence is a totality of objects (sometimes just a set, but often more than that) with respect to which the sentence is used. In ordinary discourse the domain of any sentence is most often simply the actual world. But it need not be. When we say, normally, that some horse is a jumper we take our domain to be the actual world, but when we say that some horse is winged what we say is false unless our domain is, say, the world of Greek mythology, in which case what we say is true. Any totality can be a domain. Worlds, actual, possible, mythical, imaginary, fictional, etc., are totalities. Sets are totalities. When we say that some prime is even our domain is the set of natural numbers. Usually when we use a sentence the domain is simply understood--especially when it is just the actual world. But whenever there is apt to be doubt concerning the domain we can always specify just what it is. Often a sentence can be disambiguated just by specifying the domain. Notice also that truth will be domain-relative. Many sentences are false relative to the actual world, for example, but true relative to some nonactual world.

As we said, the denotation of a term used in a sentence is a function of both its extension and the domain of the sentence. Consider the term 'horse.' All horses are included in its extension. The term 'horse' extends over not just the horses of the actual world as it now is, but also over all past and future horses of the actual world, and, importantly, horses of any other domain. Actual, former, future, possible, mythical, fictitious and imaginary horses are all included in the extension of 'horse.' Now since a term is always used in a sentence, and since a sentence is always used relative to some domain, the objects actually denoted by a term need not exhaust its extension. Generally, the denotation of a term used in a sentence will consist of just those objects in its extension which are members of the sentence's domain. Before going on it is important to note that there is a sense in which what a term signifies is semantically prior to what its extension happens to be. This is so because what can be taken to be in the extension of any term is dependent upon what that term signifies. What is in the extension of a term is just whatever is taken to have the property it signifies.

We have seen thus far that an uncharged, absolute term, a lexical item, comprehends a feature, and that a charged term, a term used in a sentence, signifies a property. Moreover, charged terms have extensions and, in the context of a sentence, denote subsets of their extensions. Can we say that an absolute term has, in any sense, an extension? The charged term 'wise' has as its extension all wise things. The charged term 'unwise' has as its extension all unwise (foolish) things. Let us say that the extension of an absolute term, e.g., /wise/, is the union of the extensions of its two charged versions. Thus the extension of /wise/ is the union of the set of wise things and the set of unwise things. We can call the extension of an absolute term a *category* (see chapter I above).

To summarize what we have seen so far, we can say that an uncharged, or absolute, term, as a lexical member outside any sentence, comprehends a feature and extends over a category. A charged term, as used in a sentence, always signifies a property and extends over all objects taken to satisfy that property. But, as every sentence is used relative to a determinate domain, a term used in a sentence denotes just those objects in its extension which happen to be included in the domain of the sentence. In what follows we will have little more to say about absolute terms, features and categories. As our interest is primarily logico-semantic our attention is on terms as they are used in sentences. We are interested in how the significance and denotation of a term is involved in the significance and denotation of sentences in which it is used.

2. There is a way of looking at the logical syntax of natural language sentences which is out of fashion among most logicians and some linguists today, but which has an old venerable history. According to this way of looking at things, sentences can be construed as pairs of charged terms connected by sentence formatives. The terms of a sentence were called "categoremata" by he scholastic logicians. The categoremata of a sentence are those expressions which were said to signify. The formatives were called "syncategoremata" and said to "cosignify" since, when attached to categormatic expressions, the results are syntactically more complex expressions which signify in their own way.

Not all formatives are sentential (i.e., sentence-forming). Some formatives simply have the effect of charging a term positively or negatively. As we saw, the formatives for negatively charging terms are explicit in natural languages (e.g., the 'un' of 'unwise,' the 'non' of 'noncommissioned,' the 'less' of 'colorless'). Positively charging formatives are rare in natural languages. Positive charge is usually tacit (but consider the 'ful' of 'hopeful' or the 'some' of 'wholesome'). But the formatives we are most interested in now are sentential. Consider the sentence 'Some officers are noncommissioned.' Here the terms, the categorematic elements, are 'officers' and 'noncommissioned.' The sentence formative is the expression 'some...are....' Natural languages, such as English, Greek and Latin, tend to split this formative in this way--part of it going with the first term and the rest going with the second term. Thus we could say that there are actually two formatives here: 'some,' which attaches to the first term to form a logical *subject*, and 'are,' which attaches to the second term to form a logical *predicate*. The first term is then the *subject-term* and the second is the *predicate-term*. The first formative, which forms subjects from terms, is called the *qualifier*. It is important to note that there is nothing about a term which in any way predetermines it to be either a subject-term or a predicate-term. Any term can occur in a sentence as either a subject-term or a predicate-term. It is this fact, that any term is fit for any term-role in a sentence, which is the foundation for syllogistic. But keep in mind that while subject-terms and predicate-terms are on a syntactical par, subjects and predicates are not.[3]

While terms *per se* are fit to play either syntactical role (subject-term or predicate-term), the interchange of terms in a sentence is not always *salva veritate* (but only *salva grammatical*). For example, we can exchange the terms in 'Every human is mortal' to get 'Every mortal is human,' but the two sentences are nonequivalent and have different truth-values. Nonetheless, there are sentences in which term interchange does preserve truth. Consider: 'Some human is mortal,' which is logically equivalent to 'Some mortal is human.' What this indicates is that the formative 'some...are/is...' is commutative or symmetric, a fact not clearly seen when the formative is split, as in English and other natural languages, into a quantifier and a qualifier. Aristotle found a solution to this by reformulating his sentences into Greek sentences which were only slightly strained, but which clearly revealed the symmetry of the 'some...are/is...' formative. He would read a sentence like 'Some human is mortal' as 'Mortal belongs to some human.' Here an expression like 'belongs to some' is unsplit and comes between the two terms--it literally connects them. And 'belongs to some' is symmetric, allowing us to commute the two terms (i.e., 'Human belongs to some mortal').

Natural language sentences consist of two terms connected by a sentence-forming functor. The formative may be split into a quantifier part and a qualifier part or it can be kept whole. This freedom is a source of great power, for it permits us to look at the logical syntax of natural language sentences in at least two different but compatible ways. How we choose to look at such sentences will depend, as we will see, upon whether our immediate interest is syntactic or semantic.

A sentence of the form 'P belongs to some S' was traditionally called an I proposition. To reveal the symmetry of the 'belongs to some' formative let us symbolize it by a multiplication sign, a paradigm arithmetical example of a symmetric formative. Thus, 'P belongs to some S' will be symbolized as 'PxS.' By introducing a symbolization for negatively charged terms we can easily symbolize an O form sentence (e.g., 'nonP belongs to some S'). We are implicitly marking the positive charge on a term with no overt sign at all. We mark the negative charge by placing a dash over the charged term. So 'nonP belongs to some S' is symbolized as '$\overline{P}$xS.'

Now terms are not the only linguistic items which are charged when in use. Sentences themselves are charged either positively or negatively. Sentences are either affirmed or denied. We say 'Some S is P' (or 'P belongs to some S') and thereby affirm P of some S. But we also say 'Some S is not P' (or 'It is not the case that some S is P' or 'It is false that some S is P' or 'P does not belong to some S' or 'No S is P' or 'Not any S is P' or 'It is not the case that P belongs to some S') and thereby deny that some S is P, deny P of some S. Notice that, as with terms, the positive charge on entire sentences is left tacit while the negative charge is explicit. Let us symbolize the negative charge on sentences in the same way we symbolized the negative charge on terms, by placing a dash over the sentence. So while the sentence 'p' is positive, '$\overline{p}$' is negative. A term and its corresponding oppositely charged term are logically *contrary*. A sentence and its opposite are logically *contradictory*.

Our ability now to symbolize negative sentences (denials) allows us to complete our symbolization of the four standard categorical forms. We are free to take an A-form sentence as the contradictory of an O-form and E-form as the contradictory of an I-form. The full schedule of "Aristotelian" forms then is

Aristotelian	I:	PxS
Forms	O:	$\overline{P}xS$
	A:	$\overline{\overline{P}xS}$
	E:	$\overline{PxS}$

These forms represent a syntax which views each categorical as a charged sentence formed by connecting (with a formative expression) two charged terms. It is possible to construct a syllogistic calculus on this syntactic foundation (the universal/particular distinction is replaced by the denial/affirmation distinction; the doctrine of distribution is preserved: terms are undistributed unless they occur under an odd number of dashes). Nonetheless, a logically more perspicuous symbolization is available. We have called the forms above "Aristotelian" in memory of Aristotle's preference for not splitting the quantitative and qualitative aspects of the term connector ('belongs to some'). Aristotelian forms are particularly useful in accounting for the semantics of assertions, statementmaking sentences, a topic to which we will soon return. For the purposes of logical reckoning, however, it is best to construct a symbolic algorithm based on a syntax which splits term connectors into subject formatives (quantifiers) and predicate formatives (qualifiers).

The "split connective" view takes each categorical sentence to consist not of two charged terms connected by a sentence formative, but as a concatenation of a subject and a predicate. Each subject is viewed as a concatenation of a quantifier and a term (the subject-term), and each predicate is viewed as a concatenation of a qualifier and a term (the predicate-term). Again, each term is charged and the entire sentence is charged. Notice that in the Aristotelian sentences, since the formative is symmetric, the order of terms is, in effect, arbitrary. There are no subjects and predicates--just two charged terms connected by a symmetric formative. On the split connective view, on the other hand, order is all-important; subjects precede predicates. Nonetheless a workable symbolization of subject-predicate sentences can be generated from the Aristotelian forms. We will call the new forms (the subject-predicate, split connective forms) "Leibnizian" in memory of Leibniz's unique contributions to the attempt to construct an algorithm for subject-predicate sentences.

The sentence-forming sign of the Aristotelian formulae can be split into a quantifier and a qualifier. The terms of I and O sentences are commutable *salva veritate*. This suggests that we symbolize the particular quantifier and the affirmative qualifier (copula) each as plus signs. In this case 'Some S is P' would be transcribed as '+S+P,' which is algebraically equivalent to '+P+S.' Again, positively charged terms can be left unmarked. But we can mark negative terms with a minus sign. So the O form, 'Some S is nonP,' is rendered as '+S+ –P,' which can be simplified algebraically to '+S–P.' The negation of entire sentences can be marked by a minus sign on the sentence as a whole so that we generate the following Leibizian schedule.

Leibnizian	I:	+S+P
Forms	O:	+S–P
	A:	–(+S–P)
	E:	–(+S+P)

These A and E forms could be further simplified to give us '–S+P' and '–S–P' respectively. A variety of reasons have been offered elsewhere for retaining the negated forms above.[4] Sommers' algorithm for syllogistic inference based on the Leibnizian schedule is particularly simple, elegant and powerful.

3. Aristotelian transcriptions are useful in revealing a perspective of sentences as charged expressions formed by connecting pairs of charged terms. They highlight the terms which are the categoremata of the sentence. Leibnizian transcriptions have the advantage of revealing the quantifying and qualifying functions of the sentence-forming connective. So we have two different but complementary ways of logically parsing a sentence. Some sentences are used to make statements. These sentences can be viewed, of course, from an Aristotelian or Leibnizian perspective. But there is a third way of viewing them. The third way is less formally perspicuous than the others. Yet it has the advantage

of revealing important aspects of the semantics of statement-making sentences in natural language. This third view, and the sentence forms generated by it, will be called 'Sommersian' in honor of Sommers' initial formulation and explication of it. The Sommersian view takes into account two important features of any sentence used in making a statement. These are: (i) Every statement-making sentence is used relative to a domain, and (ii) Every statement-making sentence is accompanied by an implicit truth-claim. We have already met (i). A sentence, whether statement-making or not, is always used relative to some determinable domain. One effect of this relativization of a sentence in use to a domain is that the denotations of terms used in it are restricted to those parts of their extension which intersect with the domain. What we want now is some understanding of (ii). We use sentences in a variety of ways to achieve a variety of ends. We use sentences in making promises and predictions, in issuing orders and warnings, in asking questions, and so forth. An important, perhaps paradigmatic, use of sentences is in statement-making, asserting. Assertions are either true or false. To make an assertion is to say something about something and, *eo ipso*, to make implicitly a truth-claim, to claim that what is asserted is true. Now Sommers' idea is that to make a truth-claim is to characterize in a certain way the domain of the statement-making sentence.

Consider a stew. We could say that it's cheap and that it's hot. 'Cheap' and 'hot' would signify two properties of the stew. We characterize the stew by affirming or denying these properties of it. A second way of characterizing the stew would be to say what its ingredients are, or are not. We could say that it has onions and carrots but no beef. We could even say that the stew is oniony and carroty but not beefy. Terms like 'oniony' ('onionish,' 'onionful,' 'carroty' and 'beefy') are used to characterize the stew in terms of its ingredients, its constituents. Not only stews but any totality could be characterized in terms of its constituents. The set of natural numbers is even-prime-ish because an even prime is one of its constituents. The world of Greek mythology is centaurish. The world of *Hamlet* is

Guildensternish. The actual world is cowish. The implicit truth-claim which accompanies a statement-making sentence is nothing more than a claim that the relative domain has some appropriate constituent. To make a truth-claim is to *constitutively characterize* the relevant domain. To say that a domain, D, is P-ish is to say that some P-thing, something denoted by 'P,' belongs to, partly constitutes, D. To say that D is un-P-ish is to say that no P-thing belongs to D. To say that D is nonP-ish is to say that some nonP-thing belongs to D. To say that D is un-nonP-ish is to say that no nonP-thing belongs to D. Note that for any D and any P, either D is P-ish or D is un-P-ish, but it need not be the case that D is either P-ish or nonP-ish. For example, the domain of natural numbers is either redish or un-redish because either it has a red constituent or it has no red constitutent. But it is neither redish nor nonredish since it has no red constituent and also no nonred (blue or pink or green or...) constituent. The domain of natural numbers is both unredish and un-nonredish.

Suppose now we make a statement by the appropriate use of the sentence 'Some officers are noncommissioned' relative to the actual world. Our implicit truth-claim then is a constitutive characterization of the actual world. Specifically, what we claim is that the actual world is noncommissioned-officer-ish, that some noncommissioned officer is a constituent of the world, that there is a noncommissioned officer in the actual world. In general then, to make a statement by the use of 'Some S is P' relative to D is to claim implicitly that D is (SP)ish. The implicit truth-claims for the four categorical forms would be, relative to domain, D,

I:	D is (SP)ish
O:	D is (SnonP)ish
A:	D is un(SnonP)ish
E:	D is un(SP)ish

Relative to a domain, D, 'P belongs to some S' is true if and only if there is an (SP)-thing in D, i.e., D is (SP)ish. An (SP)-thing being in D is a constitutive state of D, a *state* of D in which some S is P. A domain is in a constitutive state whenever it succeeds in having a constitutive characteristic. Let [p] be the state of D when the sentence 'p' relative to D is true. We will say that 'p' is true whenever [p] *obtains* and false whenever [p] does not obtain. In either case 'p' will be said to signify [p]. Equivalently we may say that 'p' expresses the *proposition* [p]. Thus the proposition that p and the state of D in which 'p' is true both refer to the same thing, viz. what is signified by 'p.' A statement-making sentence is true whenever the state or proposition it expresses (signifies) obtains. A state or proposition which obtains is a *fact*. So a statement-making sentence is true just in case it signifies a fact. Not every use of a sentence is statement-making (and thus accompanied by a truth-claim). But every sentence will signify a state (express a proposition). We can think of this as the *propositional content* of the sentence. Whether we make a statement by use of a sentence of the form 'P belongs to some S,' or ask a question or make a promise (i.e., 'Some of my children are educated,' 'Are some of my children educated?,' 'Some of my children will be educated') in each case we express the same proposition. The assertion, question and promise all have the same propositional content. What makes a use of a sentence statement-making is the understanding that it is accompanied by a tacit truth-claim. Any sentence in use expresses a proposition (signifies a state). The use of any sentence to make a statement is accompanied by the implicit claim that what is signified obtains.

We saw above that each charged term of a sentence has a signification. We have now seen that entire sentences have significations as well. Positive sentences signify states. Thus 'p' signifies [p] and, when used to make a statement, claims that [p] obtains. What of negative sentences? We are tempted to say that the statement-making use of '$\bar{p}$' ('not p,' the contradictory of 'p') claims that [$\bar{p}$] obtains. Now [$\bar{p}$] obtains if and only if [p] fails to obtain (i.e., '$\bar{p}$' is true if and only if 'p' is false). Could we say that '$\bar{p}$' signifies [p] and claim that it fails to obtain (rather than that '$\bar{p}$' signifies [$\bar{p}$] and claim that it obtains)? The problem here, of course, is that not all uses of sentences are statement-making. Our thesis is that sentences, like terms, are charged. Thus the sign of sentence negation belongs to the propositional content of the sentence. It is part of what R. M. Hare called the "phrastic" as opposed to the "neustic."[5] Frege raised the question of whether to treat sentence negation as part of the content or not in several of his essays.[6] He asked at one point: "Does negating go along with judging? Or is negation part of the thought that underlies the act of judging?" ("Negation," p. 129). In other words, is sentential negation a neustic, pragmatic element (i.e., a mode of "judging"), or is it a phrastic element, an element of the propositional content (i.e., a part of the underlying "thought")? Frege argued that the second alternative was to be preferred since it would effect an "economy of logical primitives and their expressions in language" (p. 130). Frege's was, of course, the correct choice here, but a more important reason for taking sentential negation as applying to the propositional content rather than as a mode of judging, viz. statement-making, is this: sentences themselves are terms--they are syntactically complex terms, but nonetheless they are terms. And terms in use are always charged positively or negatively. What the statement-making use of '$\bar{p}$' claims is that [$\bar{p}$] obtains. While the failure of [$\bar{p}$] to obtain may be a necessary condition for the truth of '$\bar{p}$,' it is not what '$\bar{p}$' claims when used to make a statement.

To summarize then, sentences used in statement-making are accompanied by a tacit truth-claim. There are a variety of equivalent ways of expressing the truth-claim of a statement-making sentence. For example, the sentence 'P belongs to some S' ('Some S is P'), when used to make a statement relative to a domain, D, claims

(a) *D is (SP)ish*, or equivalently,
(b) *an (SP)-thing is a constituent of D*, or equivalently,
(c) *the proposition expressed by 'p belongs to some S' (viz. [PxS]) obtains*, or equivalently,
(d) *the state signified by 'P belongs to some S' (viz. [PxS]) obtains.*

To say that 'PxS' is true is simply to say that the claim implicitly made by the statement-making use of that sentence relative to a specified domain is satisfied, that [PxS] is a fact.[7]

4. Just as a term outside of its use in a sentence neither signifies nor (on the basis of what it signifies) denotes, a sentence outside of its use in any discourse context neither makes a statement nor fails to make a statement. Contextual elements determine whether or not the use of a sentence is statement-making. If it is, then these elements indicate that the sentence is understood as being accompanied by a truth-claim. Let us use Frege's assertion sign as an explicit mark of the neustic element indicating statement-making. Then when we say 'Some officers are noncommissioned' in a statementmaking context we could indicate this by the formula '$\vdash(\overline{C}xO)$' (or '$\vdash(+O-C)$'). Here '$\vdash$' merely indicates the statement-making force of the sentence which it precedes. We could even think of it as a short-hand for the tacit truth-claim which accompanies the sentence in statement-making uses.

Now we have seen that terms in use are charged, e.g., 'P'/'$\overline{P}$,' 'commissioned'/'noncommissioned,' and that sentences as well are charged, e.g., 'p'/'$\overline{p}$,' 'Some man is wise'/'No man is wise.' Does '$\vdash$' represent an additional charge on statement-making sentences? Could it be a kind of "assertoric charge," correlated with a negative charge (perhaps '$\nvdash$')? We might think that any sentence can either be asserted (used to make a statement) or rejected (used to withdraw or reject a statement). But this would be to confuse the assertoric force of a sentence in its statement-making use with its logical charge. Sentences are positive or negative in charge as they are affirmed or denied. Affirmation (e.g., 'P belongs to some S') and denial (e.g., 'P belongs to no S') are polar opposites (allowing us to represent them as opposing charges on sentences). But asserting is not a charge on sentences at all. Every sentence used in making a statement is asserted. What is correlative to asserting, or statement-making, is not any negative-looking element like rejection, but simply other things like asking, promising, warning, ordering, etc.

Sentences then, like terms, are either positively or negatively charged. In other words, while terms come in contrary pairs, sentences come in contradictory pairs. Now we saw above that sentences can be thought of as complex terms. It is this fact which allows us to bring together the various claims made thus far. Since any term in use has a signification and a denotation (determined by the domain and the term's extension), it follows that sentences in use must as well have signification and denotation. We have already seen that what a sentence signifies is a state or proposition (which can be explicated in terms of a constitutive characteristic of the domain relative to which the sentence is used). What does a sentence denote?

Every used term, whether syntactically simple or complex, both signifies and denotes. What is signified is always a property or, in the case of sentences (sentential terms), constitutive characteristics (special kinds of properties). What is denoted are objects (having that property). Again it should be remembered that a term's denotation is just that part of its extension falling within D. In other words, what a term denotes is always the object or objects in the domain which satisfy the property it signifies (i.e., the intersection of its extension and the domain). Since a sentence signifies a state (proposition, constitutive characteristic), its extension is what satisfies that characteristic, what is in that state, viz. the set of domains in that state. Let '$\underline{p}$' indicate the extension of 'p,' i.e., the set of domains characterized by [p]. What a statement-making sentence denotes, then, is the intersection of this set and the domain relative to which the statement is made. Since either the domain relative to which a sentence is used to make a statement is included in the set of domains characterized by the (constitutive) property signified by the sentence or it is not, the relevant intersection (of the relevant domain and the set of domains characterized by the constitutive characteristic signified by the sentence) will either be equivalent to that domain or be empty. Finally, we say that an assertion (statement-making sentence) whose denotation is empty is false; otherwise it is true.[8]

Let '...T...' D be a term used in a statement made relative to domain D. Let T̲ be its extension, '|' abbreviate "signifies" and '↓' abbreviate "denotes." Then, generally,

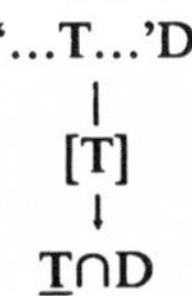

More specifically, when the term is sentential:

'p'D
|
[p]
↓
p̲∩D

where p∩D is equivalent to either D or the null set. Notice that if we use a sentence to make a statement then our implicit truth-claim is simply that what the sentence denotes is in the state it signifies. A statement-making sentence is true if and only if the domain it denotes is in the state (has the constitutive characteristic) it signifies, i.e., the proposition it expresses obtains, is a fact. A true statement-making sentence, then, denotes its domain; a false one denotes nothing.

We consider now the following examples, where a is the actual world and g is the world of Greek mythology.

1. 'Some logicians are mathematicians' asserted relative to the actual world.

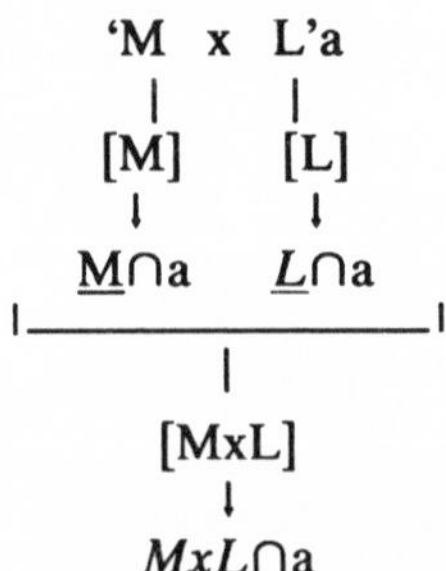

This shows us that 'logician' denotes actual logicians, 'mathematician' denotes actual mathematicians, the entire proposition denotes the intersection of the set of domains containing mathematical logicians (logicians who are mathematicians) and the actual world. Since the actual world contains logicians who are mathematicians, the relevant intersection is nonempty--the assertion is true.

2. 'All philosophers are literate' (='No philosopher is illiterate') relative to the actual world.

$$\overline{\text{'}\bar{L}\ x\ P\text{'}a}$$

$$[\bar{L}] \quad [P]$$

$$\downarrow \qquad \downarrow$$

$$\underline{\bar{L}}\cap a \quad \underline{P}\cap a$$

$$[\overline{\bar{L}xP}]$$

$$\downarrow$$

$$\underline{\overline{\bar{L}xP}}\cap a$$

This shows that 'philosopher' denotes actual philosophers, 'illiterate' denotes actual illiterates, the entire sentence denotes the intersection of the set of domains not characterized by the presence in them (i.e., characterized by the absence from them) of illiterate philosophers and the actual world. Since the actual world (presumably) contains no illiterate philosophers, the assertion is true.

3. 'Some horse flies' asserted relative to the actual world.

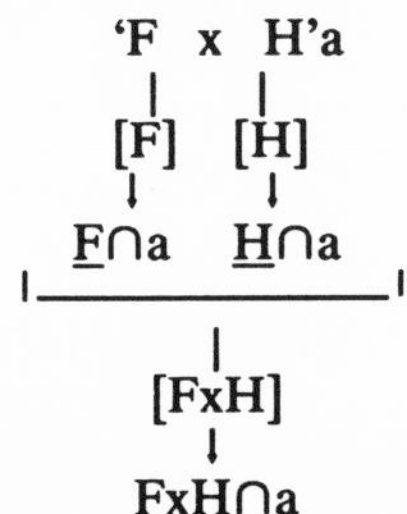

Since the actual world is not a domain containing flying horses, the denotation of this sentence is the empty set. The assertion is false. We get a true statement when a is replaced by g.

4. 'Nothing on Pluto is alive' asserted relative to the actual world.

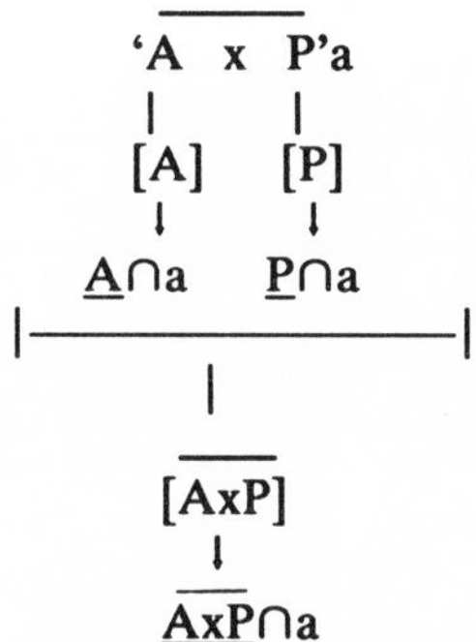

Here 'P' denotes actual things on Pluto, 'A' denotes whatever is actually alive, the entire sentence denotes the intersection of the actual world and the set of domains which do not contain any live things on Pluto. Since, at this time, we do not know whether the actual world is in this set, we do not know whether $\overline{\underline{AxP}}\cap a$ is empty or not--the truth or falsity of the assertion is unknown.

5. 'Some man loves a woman' asserted relative to a.

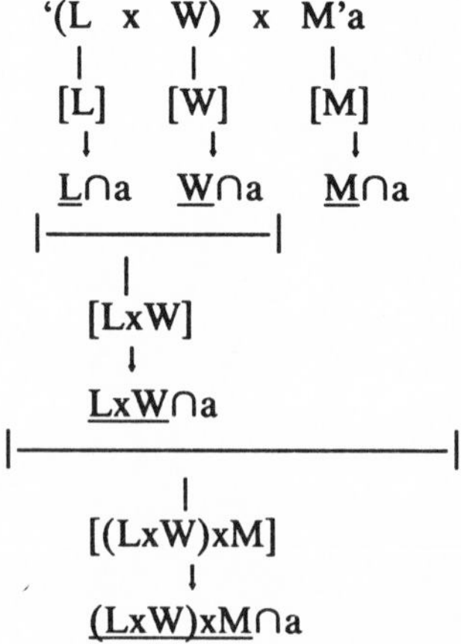

Here 'loves' denotes actual lovers, 'loves a woman' denotes actual woman-lovers (lovers of some woman) and the entire sentence is true just in case the actual world is among the set of domains containing a man who is a woman-lover.

6. 'If p then q' asserted relative to a specifiable domain D.

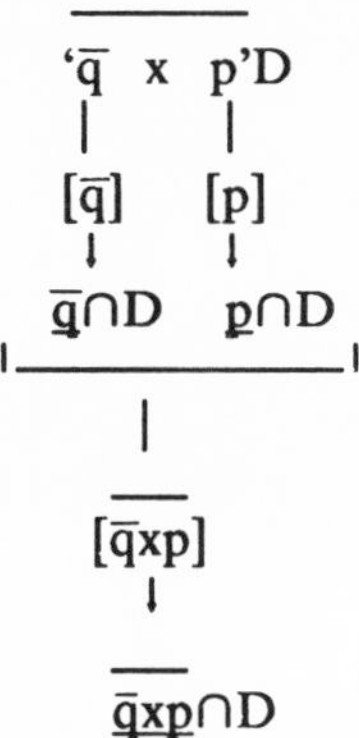

In this example 'p' and '$\bar{q}$' signify constitutive properties and denote the intersections of the sets of domains having each such property respectively and D. The entire sentence denotes the intersection of D and the set of domains which are not characterized by both the constitutive properties signified by 'p' and '$\bar{q}$.' In other words, the sentence denotes the intersection of D and the set of domains in which 'p' and '$\bar{q}$' are both false. Thus, whenever 'p' is true and 'q' is false (i.e., $\underline{p}\cap D = D$ and $\underline{q}\cap D = \phi$) 'if p then q' will be false, since in such a case $\overline{\underline{q}x\underline{p}}\cap D = \phi\cap D\cap D = D$ (i.e., its contradictory is true).

5. When *used* in a sentence a term like 'red' denotes all the red things in its extension which are also in the domain of the sentence in which it is used. When *mentioned* in a sentence, however, such a term denotes tokens of its use. Thus, in using a sentence like "'Red' is three-lettered" the term 'red' is mentioned rather than used. What is used is the name, or quotation, of 'red,' and what it denotes are tokens of the use of 'red'--tokens of 'red.' What the name of 'red' signifies is simply the property of being a token of the use of 'red.'

When we say 'Socrates is wise' we characterize Socrates as having the property of wisdom (i.e., [wise]). But sometimes we want to say things about properties, such as wisdom, themselves. We say, for example, 'Wisdom is rare.' Here what is being referred to is the property of wisdom--not wise things. In effect our sentence says '[wise] is rare.' While 'wise' signifies [wise] and denotes what has [wise] (i.e., wise things), '[wise]' signifies the property of being wisdom (i.e., [[wise]] or [wisdom]) and denotes what has that property, which, since the only thing which has or could have the property of being wisdom is just the property of wisdom, is the property of wisdom itself (i.e., [wise]). In short, then, 'wisdom' signifies the property of being wisdom ([wisdom], [[wise]]), and denotes wisdom ([wise]), which in turn happens to be signified by 'wise.'[9] By analogy 'John' in 'John is tall' signifies the property of being John ([John]) and denotes John.

As we have seen, it is important to distinguish between, say, the term 'wise,' the name of the term 'wise,' and the term 'wisdom.' It is equally important to distinguish between a sentence, the name of that sentence, and the name of the state or proposition signified or expressed by that sentence. Just as we can use a sentence in which we mention rather than use a term (viz. by using the name of that term) we can use a sentence in which we mention another sentence. For example, we might say "'Socrates is wise' is English." Here the term 'Socrates is wise' denotes the tokens of the sentence 'Socrates is wise' and signifies the property of being a token of 'Socrates is wise' (i.e., ['Socrates is wise']).

6. In this section we offer definitions for the notions of *logical entailment* and *logical truth*.

> An assertion A_1 logically entails an assertion A_2 if in any domain where A_1 is true A_2 is true (equivalently, the set of domains in which A_1 is true is included in the set of domains in which A_2 is true; equivalently again, the extension of A_1 is included in the extension of A_2).

Using familiar notation, this amounts to

$$A_1 \vdash A_2 \text{ iff } \underline{A}_1 \subseteq \underline{A}_2$$

We define logical truth as follows:

> An assertion A is logically true iff A is true in every domain (equivalently, the extension of A is the set of all domains).

Letting '$\cap$' stand for the set of all domains, this amounts to

$$\vdash A \text{ iff } \underline{A} = \cap$$

Our semantic theory also allows us to account for a variety of other semantic notions. For example, we can easily account for the distinction between *synonymy* and *coextensiveness.* If two terms are synonymous they signify the same property (and thus are coextensive as well). For example, 'wed' and 'married' commonly signify a property, and thus extend over the same things. But, to use a Quinean example, 'renate' and 'chordate,' even though sharing a common extension (viz. all normal vertebrates), are not synonymous, since each signifies a different property (viz. the property of having a kidney and the property of having a heart, respectively).

Just as a pair of terms can share a common extension and, sometimes, a common signification as well, a single term can have more than one signification (and consequently more than one extension). Such terms are *semantically ambiguous.* For example, 'bank' is ambiguous in (at least) two ways. It signifies the property of being a river's edge and the property of being a certain kind of financial institution. Likewise, 'painting' is ambiguous. It signifies the property of being a certain kind of work of art and it signifies the different property of being an activity involving the coloring of surfaces.

The following figure illustrates the semantic notions just examined (the arrow here stands for 'extends over').

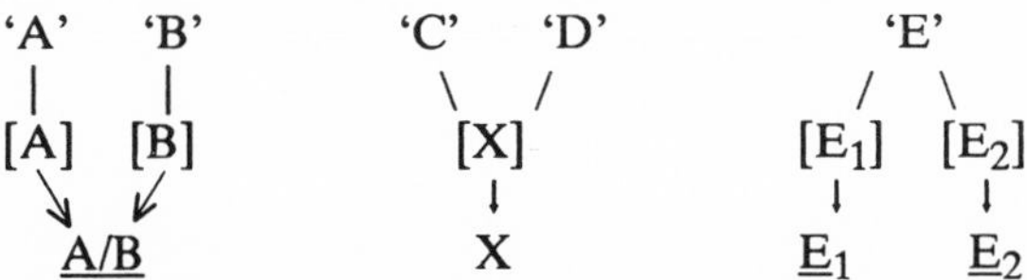

Terms 'A' and 'B' are coextensive, 'C' and 'D' are synonymous and 'E' is semantically ambiguous.

Finally, we can account for the ambiguity and synonymy of entire sentences. A sentence is ambiguous whenever it signifies more than one constitutive property. For example, any sentence which uses one or more semantically ambiguous terms will itself be (semantically) ambiguous. Also, a sentence which has more than one formal analysis (e.g., 'Some authors don't finish every book') will be (syntactically) ambiguous. Two sentences are synonymous whenever both signify the same constitutive property. Thus, 'Socrates is tired' (uttered by Plato) and 'I am tired' (uttered by Socrates) are synonymous. They both signify the constitutive property of being tired-Socrates-ish, of having as a constituent a tired Socrates. Again, Socrates utters synonymous sentences when he utters 'I am tired' and 'Je suis fatiqué.' Both sentences signify the same constitutive property.

Note that what we have said here concerns sentences--not assertions. An assertion is a sentence uttered in order to make a truth-claim (as opposed to a sentence uttered in order to ask a question, make a demand, offer a prayer, make a promise, etc.). An assertion (i.e., a sentence used to make a truth-claim) is always used relative to a specifiable domain. Thus its denotation is specifiable. Sentential synonymy and ambiguity, however, can be accounted for in terms of signification alone, without recourse to denotation. At this point it might be helpful to note that if two expressions are synonymous (have the same signification, signify the same property) they have the same extension. If they have the same extension they have the same denotation (when used in statement-making). If they have the same denotation they make the same truth-claim (if they are sentential terms). If they make the same truthclaim they have the same truth-value. The converses of none of these hold.

Chapter V

Endnotes

1. A slightly different account of our semantic theory is found in Englebretsen (1985f).

2. See Englebretsen (1987b).

3. For further details on this important distinction, see Englebretsen (1981d) and (1985a).

4. These are found in Chapter 14 of Sommers (1982) and in Chapter VI below.

5. Hare (1964), pp. 17-18.

6. See especially "Negation" in Frege (1952), pp. 117-135, and Frege (1979), pp. 185 and 196.

7. We note in passing here that one of the main advantages Sommers has claimed for this view of statement-making sentences in terms of domains and constitutive characteristics is that it permits the formulation of a correspondence theory of truth which is not subject to Strawson's legitimate concern to find the objective correlates (states, facts) of true sentences. The correlates are not mysterious states or facts (or even existences) but just (constitutive) characteristics of the domain. See Sommers (1987).

8. One is tempted to see this as support for the inclination, especially among nineteenth-century British logicians, to use '1' for both the universe of discourse and 'true' and '0' for both the null set and 'false.'

9. Similarly, in Mill (1943) (especially I.ii.5), Mill held that an abstract name such as 'wisdom' denotes just the attributes (properties) which the corresponding general term, 'wise,' connotes (signifies).

Chapter VI

QUADRATUM AUCTUM*

There is a commonly-held view that the logical relations exhibited by a square of opposition can only be so displayed for a limited class of assertoric sentences. On this view there are certain kinds of such sentences that find no place on the square. Until very recently I too held this view and have defended versions of it publicly on several occasions.[1] Nevertheless, I propose here to reject that view. Instead, I want to show that a close inspection of the features of the square reveal a rarely realized, or even suspected, power and range. To be sure, the square I speak of is not exactly the traditional square of opposition. What I have in mind is an amended and augmented version of the old square. In almost every case these new features of the square are the result of distinctions which have always or usually been overlooked by modern logicians. Most of them are due to Fred Sommers, and I take this opportunity to pay homage to his keen, clear logical vision. The paths I follow here are usually ones he has discovered. Sommers has produced a steady flow of logical studies during the last two decades.[2] But it is his new book, *The Logic of Natural Language*,[3] especially chapter 14, which is the strongest and most recent inspiration for my following remarks.

1. What is the most basic device available for making reference? In other words, how is initial reference made? Now there are, of course, a variety of proffered answers. For some the claim is that reference is initially and fundamentally achieved by the use of proper names. For others reference is so achieved by the use of "logically proper names." For still others it is the result of the use of an indefinite description. Without arguing much for it,[4] I want to hold that initial, primary reference is made, either explicitly or implicitly, by the use of indefinite descriptive phrases such as 'a man,' 'an egg,' 'someone,' 'something,' 'some girl,' etc.[5]

I shall also accept without arguing for it a thesis suggested by Aristotle and advocated by Leibniz. It is the claim that all assertoric sentences can be viewed as, in some way, logically categorical, consisting of a single subject and a single predicate.[6]

Finally, I accept without arguing (because it has already been argued for extensively by Sommers) the important distinctions between denial and predicate negation and between denial and sentential negation. One can deny a predicate of a given subject as well as affirm that predicate of that subject. Moreover, one can affirm or deny the negation of that predicate of that subject. Contradiction is the relation between two sentences sharing a common subject and a common predicate, but such that the predicate is affirmed in one case and denied in the other. This contrasts with the modern view which takes contradiction to be a relation between a sentence and its negation (achieved by the application of a sentence function--sentential negation).[7]

Given these brief preliminary remarks, let us say that a *primary sentence* is any assertoric sentence whose subject is a phrase which can be used to make an initial reference, or the contradictory of such a sentence. Sentences like 'a man is coming,' 'some girl is teasing Jim,' and 'someone isn't being honest' are primary. In traditional terminology each of these is an I or O proposition. Each is a particular affirmative or particular negative. Note that what is negative about O propositions is the predicate. Thus, 'someone isn't being honest' could be rephrased as 'someone is being dishonest.' Sentences of the form 'some S is P' and 'some S isn't P' (='some S is nonP') are both affirmations. The O form simply affirms the negation of the predicate affirmed in the corresponding I form. So sentences of the I and O forms are primary. Their contradictories are primary as well. The contradictory of an affirmation is a denial. Consider 'a man is coming.' How is this denied in English? Sometimes by prefixing a negator (e.g., 'not a man is coming,' 'not a creature was stirring'). Thus, the modern logician is easily led to believe that contradiction is achieved here by the application of sentential negation. But the 'not' here does not negate a sentence. It indicates that the predicate, 'is coming,' is being denied of the subject, 'a man.' Most often we contract 'not a' to get 'no man is coming.' So if 'an S is P' and 'an S isn't P' are primary, then their contradictories, 'no S is P' and 'no S isn't P,' are primary as well. This suggests the following *primary square of opposition*.

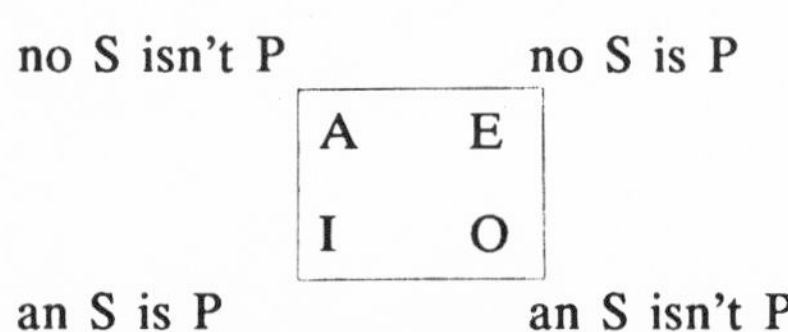

Notice that I've taken the liberty of labeling the A and E sentences. 'No S is P' is the usual formulation for an E sentence, but the labeling of 'no S isn't P' as A is admittedly unusual.[8] One misses the universal quantifier here. I can only say for now that the universal quantifier is not far off, and besides, I think we can learn to love my A form.

Taking the I and O forms ('an S is P,' 'an S isn't P') as primary (in part because they represent our initial modes of reference), the respective A and E forms are simply generated by forming the contradictories of I and O. The purpose of forming any square of opposition is to display the logical relations which hold among an intimate group of syntactically close sentences. These relations are dictated by the rules which govern the construction of the square. The first, and most obvious rule governing the primary square is the *law of excluded middle.*

(LEM) Either a sentence or its contradictory is true.

In terms of our square LEM says that either I or E is true and either O or A is true. LEM immediately calls for a second rule which excludes the possibility of both a sentence and its contradictory being true. This is the *law of noncontradiction.*

(LNC) A sentence and its contradictory are not both true.

The primary square is constructed in accordance with LEM and LNC. No other law determines the primary square.

I want now to augment the square by adding to it certain elements in what will appear as shocking arrangements. I begin with universally quantified sentences. We must find a place on our square for sentences of the forms 'every S is P' and 'every S is nonP.' I will call these the a and e forms, respectively. Where do a and e belong on the square? It is clearly not possible for an S to be P while every S is nonP, nor for an S to be nonP while every S is P. This suggests that I and e, as well and O and a, form opposite pairs. Now LEM and LNC will rule here only on the condition that A=a and E=e. I will call the law which governs the oppositions of I/e and O/a the *law of quantified opposition*, where two affirmations are quantified opposites when their subjects differ only in quantity and the predicate of one is the negation of the predicate of the other.

(LQO) A sentence and its quantified opposite are not both true.

What LQO says is that I and e are not both true and that O and a are not both true. We can add these new features to the primary square to get this *augmented square of opposition.*

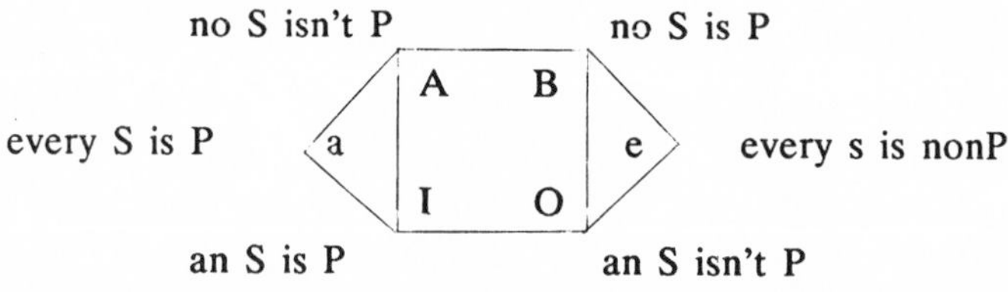

The quantified opposite of a sentence will be its contradictory (i.e., the above square will simply collapse into the primary square) only if A=a and E=e. Now it certainly is the case that a logically implies A and that e logically implies E. For example, if every god is immortal then no god is mortal. But does A imply a, and does E imply e? I want to show later that such implications do not always hold. Until then just consider the two sentences 'no unicorn is unridden by me' and 'every unicorn is ridden by me.'

It is tempting to assume that, given any subject, any predicate is such that either it or its negation will hold of that subject.[9] Let us call two sentences which differ *only* in that the predicate of one is the negation of the predicate of the other *logical contraries*. I and O are logical contraries; so are a and e. We are tempted, then, to adopt the *predicative law of excluded middle.*

(PLEM) Either a sentence or its logical contrary is true.

A companion law says that a and e cannot both be true. This law undoubtedly holds and was recognized as early as Aristotle.[10] We can call it the law of incompatibility since logical contraries are incompatible).

(LIC) Logically contrary universal sentences are not both true.

While LIC says that a and e are not both true, PLEM says that one of them is true, and also that I or O is true. Now there are no reasons to reject LIC. But there are times when PLEM does not hold. In other words, there are sentences which are not true in their I, O, a, or e forms. Such sentences are vacuous.[11]

2. At *On Interpretation* 21a26-28 (and again at Categories 13b14-19) Aristotle argued that if Socrates exists then either 'Socrates is ill' or 'Socrates is well (= nonill)' is true, but that if Socrates doesn't exist neither are true. In other words, PLEM fails when the subject fails to refer. Here 'Socrates is ill' and 'Socrates is nonill' are logically contrary. PLEM holds (i.e., one or the other is true) only as long as 'Socrates' refers successfully. Let us say that a vacuous sentence is

one for which PLEM fails. One kind of vacuousity is due to the failure of reference by the subject. But when does such reference fail? Whenever what is purportedly referred to by the subject doesn't exist? Usually, but not always.

Sometimes when I produce 'an S is P' my audience is justified in concluding 'an S exists.' Not because 'an S is P' logically entails 'an S exists' (contra Russell), for it doesn't. Nor because 'an S is P' presupposes 'an S exists' (contra Strawson), for it doesn't. Were the Russellian point correct there would be an argument of the form

some A is B
therefore, some A is C

which is invalid. Were the Strawsonian point correct there would be an argument of the form

some A is B
some A is C
therefore, some A is C

which begs the question (note that what is presupposed has the status of a suppressed premise). How can my audience validly and without begging any questions draw the conclusion 'an S exists' from my 'an S is P'? Only by admitting a suppressed premise of the form 'every P exists.' The argument scheme then is

an S is P
every P exists
therefore, an S exists

a Darii syllogism.[12]

Let us suppose now that the conclusion, 'an S exists,' does not hold. Then either 'every P exists' is not a suppressed premise, or 'an S is P' is false. In our normal discourse we take our references to succeed. We presuppose that every thing of which we speak exists. I say 'a man once walked on the moon,' presupposing that whatever has walked on the moon exists, and thus that he, being one of them, exists. Such discourse is *factual*. When we produce 'an S is P' in a factual discourse situation we presuppose 'every P exists,' and thus imply 'an S exists.' It is the presupposition which qualifies our discourse as factual. So, in a factual discourse situation, whenever 'an S exists' is false, so is 'an S is P.' But sometimes our discourse is not factual but *fictitious*.[13]

In a fictitious discourse situation 'an S exists' is false but 'an S is P' may still be true since the presupposition, 'every P exists,' is not made. Clearly, failure of reference results in vacuousity only in factual discourse situations. You would misunderstand me if you concluded from my 'a man called Santa lives at the North Pole' that a man called Santa exists. For the conclusion requires the hidden premise that whoever lives at the North Pole exists. I haven't got that presupposition because my discourse here is (normally) fictitious. Were the situation not fictitious but factual (suppose I'm a young, gullible child), then the fact that a man called Santa does not exist would indeed render my sentence false.

In general, then, for factual discourse, if there is no S then both 'an S is P' and 'an S isn't P' will be false--PLEM will fail. What of 'every S is P' and 'every S is nonP'? When the subject is empty (i.e., both I and O are false) a and e will be undefined (in effect, a is defined as the conjunction of A and I while e is defined as the conjunction of E and O). That is why we can say 'a unicorn is in my house' is false, it's contradictory, 'no unicorn is in my house,' is true, but cannot say 'every unicorn is out of my house' (= 'every unicorn is non-(in my house)').

Failure of existence for referents in factual discourse is only one source of vacuousity (i.e., PLEM failure). PLEM also fails to hold whenever a subject is undetermined with respect to its predicate. Suppose I say 'a man will walk on Venus in 2190.' Maybe. We just don't know yet. The subject is undetermined (indeed, undetermined for now) with respect to the predicate. We do know now that either a man will walk on Venus in 2190 or no man will (i.e., LEM holds), but we don't know which. We are following Sommers in saying that sentences of the form 'every S is P' and 'every S is nonP' (a and e) are undefined whenever I and O are both false (as when, in factual discourse, 'an S' fails to refer) or whenever I and O are undetermined with respect to truth-value (as when) 'an S' is undetermined with respect to 'P' and 'nonP'). In either case PLEM will fail to hold.

Suppose PLEM does hold. Then either I or O is true, and either a or e is true. In fact, in such cases A=a and E=e. For such nonvacuous cases, in other words, A will imply a and E will imply e (the converses always hold). To see this notice, for example, that if A is true O is false (LNC); if O is false then I is true (PLEM); if I is true then e is false (LQO); if e is false then a is true (PLEM); therefore, if A is true a is true. So for nonvacuous cases we could simplify the augmented square to give us

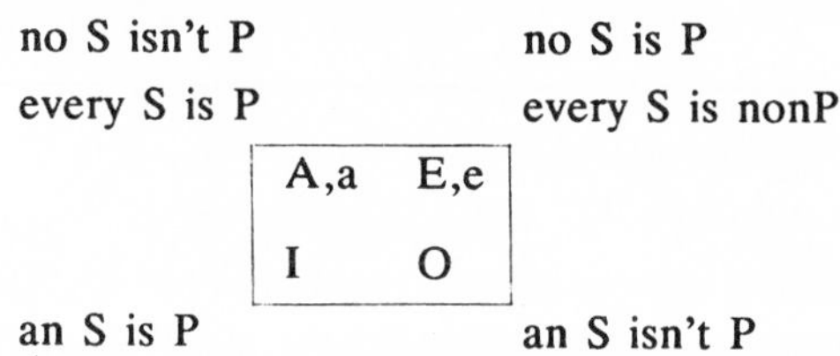

The traditional rule of obversion (i.e., A=a, E=e) obtains only for this square, i.e. only for nonvacuous, PLEM-governed sentences. One version of the simplified square is the *traditional square.*

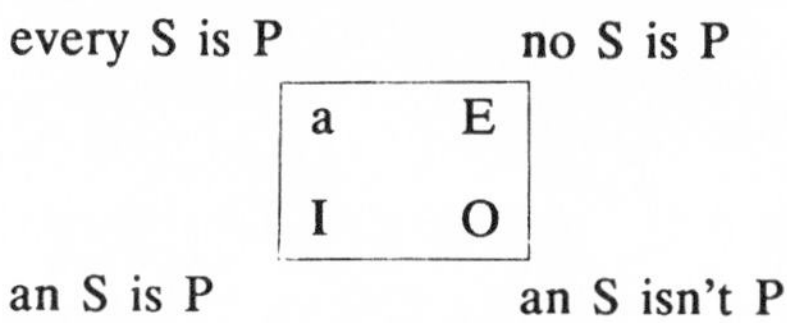

Consider now an S (say an egg). Is it P or nonP? To some this question seems very odd. To others it seems in perfect order. For the latter, no matter what P is, S is P or nonP. Our egg is green or nongreen (O.K.); it is round or nonround (O.K.); it is happy or unhappy (??); it is patriotic or nonpatriotic (!); it is poetic or nonpoetic (!!). In a series of studies beginning in 1959 Sommers[14] attempted to establish and reinforce the notion that not every predicate is appropriately, sensibly predicable of every subject.[15] Some things *can* be sensibly said of our egg, others cannot. Notice that what can be sensibly said of a thing need not be true of it. Thus, we can sensibly say of our egg that it is green. In general if 'P' is predicable (can be sensibly predicated) of a given subject then so can any term contrary to 'P.'[16] For example, 'green' is predicable of whatever 'red,' 'yellow,'

'blue,' 'pink,' etc., are predicable. The *logical* contrary of a term can be do fined as the disjunction of all of its contraries. So 'non-green'='red or yellow or blue or pink or....' This means, in effect, that when considering predicability we can ignore the distinction between a term and its negation (logical contrary). 'P' and 'nonP' are always predicable of the same things. Sommers notation '/P/' (reminding us of the mathematician's notion of an absolute number) can be used to indicate either 'P' or 'nonP.' '/P/' is read 'absolute P.' If an S is P then it is /P/ (the converse does not generally hold). Likewise, if S is nonP it is /P/.

Suppose now that we have an S which is not /P/ (e.g., our egg, which is not /poetic/). If an 'S is /P/' is false then so are 'an S is P' and 'an S isn't P' (= 'an S is nonP'). So PLEM will not hold--the sentences are vacuous. In other words, a sentence will be vacuous when its predictable of its subject. Such sentences are often classed as category mistakes; they are senseless. Category mistakes, like sentence with empty subjects and sentences with subjects undetermined with respect to their predicates, are vacuous, false in their I and O forms--PLEM does not hold.

We can summarize what has thus far been established by the following augmented square, to which we add arrows to indicate valid implications.

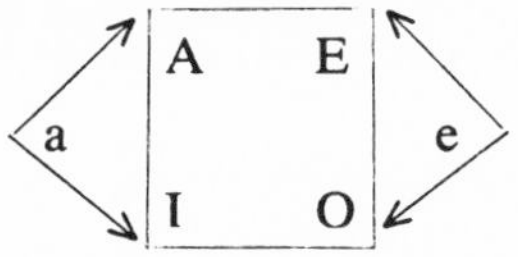

A E
a e
I O

For nonvacuous cases we can add arrows from A to a and from E to e.

3. Aristotle generally ignored singular subject.[17] The scholastic tended to take sentences with singular subjects to be implicitly universal. The view adopted in the present study is that all assertoric categoricals consist of a subject and a predicate. Moreover, a subject is a syntactical complex, consisting of a term and a quantifier. Reference is the role of subjects, and is achieved by the combined efforts of the quantifier and the denotation of the subject term. Let us

say that the denotation of 'logician' is Aristotle, Leibniz, Frege, Russell.... The universal subject, 'every logician,' refers to the entire denotation of 'logician,' i.e., Aristotle, and Leibniz, and Frege, and Russell, or.... The particular subject, 'some logician,' refers to an undetermined (perhaps determinable part, perhaps all) of the denotation of 'logician,' i.e., Aristotle, or Leibniz, or Frege, or Russell, or...(with inclusive 'or'). Scholastics took singular subjects to be implicitly universal since they refer to all of their denotations. Thus 'Socrates' denotes Socrates and refers (when in the role of logical subject) to all of that denotation, i.e., Socrates. Later Leibniz[18] and still later Sommers[19] discovered that singular subjects could be taken as implicitly either universal or particular--arbitrarily. For the reference of 'some Socrates' is just a part of the denotation of 'Socrates,' which, since it has but one part, is again just Socrates. So 'every Socrates'='some Socrates'='Socrates.'

The fact that singular subjects have this "wild" (Sommers) quantity has interesting consequences for the logical relations displayed by a square of opposition. The singularity of the subject can be manifested by recognizing the implications from the I form to the a form and from the 0 form to the e form. Thus:

Indeed, this could be simplified as a *singular square*.

A	E
I,a	O,e

In the case of vacuous singular sentences the I,a and O,e forms will be false and their contradictories, A and F, will be true.

4. I want to summarize at this point the features of the augumented square of opposition and then offer some sample sentences. The augmented square is arrived at by adding to the primary square.

Primary Square

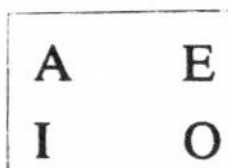

The primary square is governed by LEM and LNC. The augmented square adds positions for universally quantified forms, a and e. Keep in mind that primary reference is achieved by indefinite, particularly quantified, or singular terms. Reference by universally quantified terms must be considered secondary. Indeed, the universal forms are defined in terms of the primary sentence forms.

Augmented Square

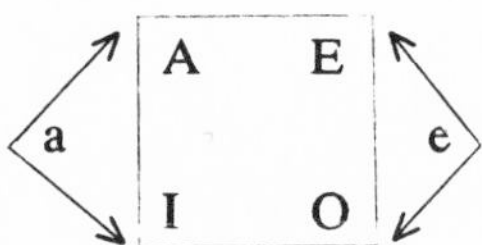

LEM and LNC still apply. LQO and LIC also hold. Whether the law PLEM now holds depends upon what further nonlogical, extrasyntactical information we have. PLEM will hold only if the subject is nonempty, determined with respect to the predicate, and the predicate is predicable of it. Otherwise, PLEM does not hold--the sentences are vacuous and false in their I and O forms. When PLEM holds (non-vacuous cases) arrows of implication from A to a and from E to e can be added. Or, alternatively, we can produce the simplified,

Nonvacuous

A,a	E,e
I	O

Knowledge that the subject is singular is an extra bit of semantic information that can also be displayed on the square by adding implication arrows from I to a and from O to c, or simply

Singular Square

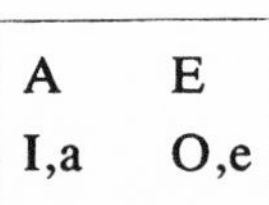

Finally notice that if we know that our sentences are both singular and nonvacuous we can add arrows from A to a, E to e, I to a, and O to e. This amounts to combining the nonvacuous and singular squares, which simply collapse together into the

Singular, Nonvacuous 'Square'

A,I,a——E,O,e

Note that these various simplifications of the augmented square always require additional, extralogical information. In the absence of any such information the augmented square displays all we know of the logical relations which hold among the various sentence forms.

Now some examples. Consider the sentence 'an S is philosophizing.' We have no knowledge of 'S' here, so the most we can offer is an augmented square.

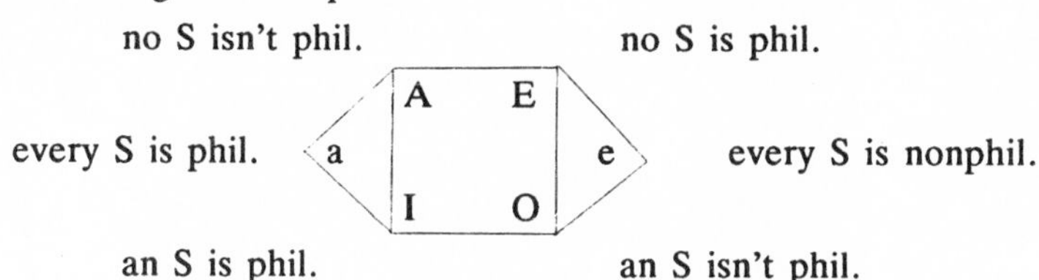

Let 'S' be 'snake.' Since 'philosophizing' makes no sense of (is impredicable of) 'snake,' 'a snake is /philosophizing/' is false. Thus, the original sentence is vacuous. It is false in its I and O forms. The case will likewise be vacuous when 'S' is 'Saturn-dweller,' since here the subject is empty. Likewise for 'S' as 'Santa's elf' since the elves which help Santa are undetermined, in the stories and tales, with respect to their intellectual activities. For such vacuous cases we can use an augmented square with truth-values added to indicate the effects of vacuousity ('X' stands for 'undefined').

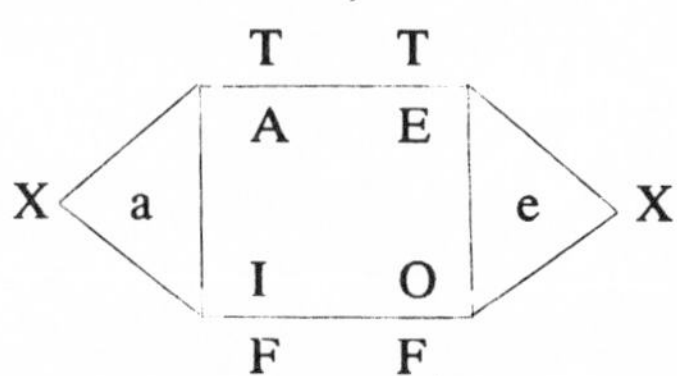

Now let 'S' be 'scientist'--a clearly nonvacuous case.

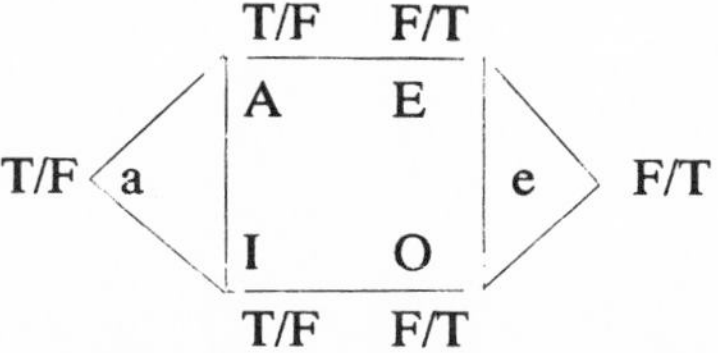

Suppose 'S' is a singular term which is empty (say 'daughter of Prince Charles') or impredicable by 'philosophizing' (say '6'), or undetermined with respect to 'philosophizing' (say 'Sinbad'). Then we have

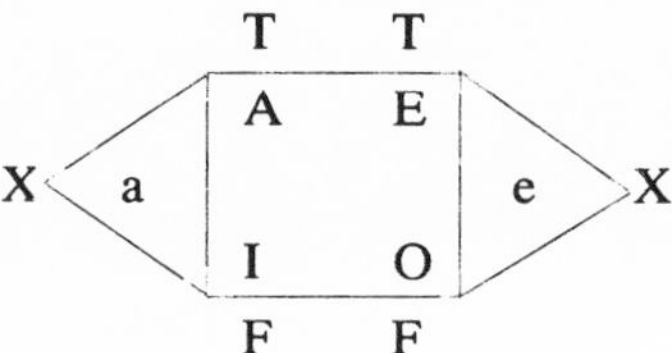

Finally, Let 'S' be 'Strawson.'

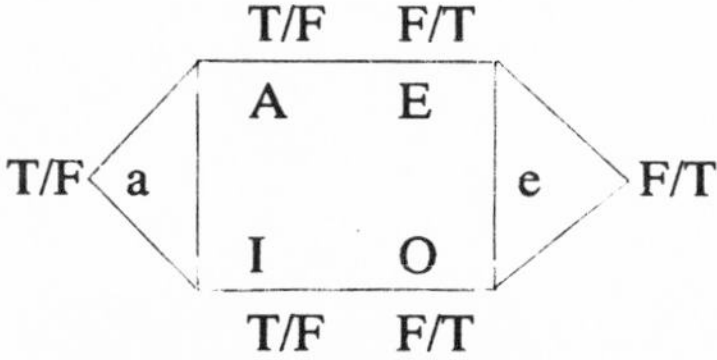

5. In this final section I want to show that even the so-called truth-functionals (compound sentences) find their place on the square of opposition. Moreover, the discovery of square relations among compounds reveals further important and interesting parallels between categoricals and compounds. Not the least of these is the one between the paradoxes of material implication and the paradoxes of existential import.

The old view was that compound sentences could be reduced to categorical forms.[20] Kant held the view that categoricals and compounds are completely separate and irreducible to one another. And contemporary logicians generally take categoricals to be reducible to compounds since they must be parsed in terms of conditionals, conjunctions, etc. Sommers has opted[21] for a view once held by Pierce.[22] According to this view, categoricals and compounds are mutually independent but share a common underlying formal structure. The result, for Sommers, is that one can build an algorithm which can be used to analyze inferences involving either categoricals or compounds. Compound forms are not reducible to or from categorical forms, but they are isomorphic with those forms. This isomorphism is sufficient to permit a single algebra to model inferences involving either categoricals or compounds. Thus a conjunction is syntactically isomorphic to a particular, while a conditional is isomorphic to a universal. Letting 'p' and 'q' abbreviate two different sentences, we can construct an augmented square with compound sentences replacing their corresponding categoricals.

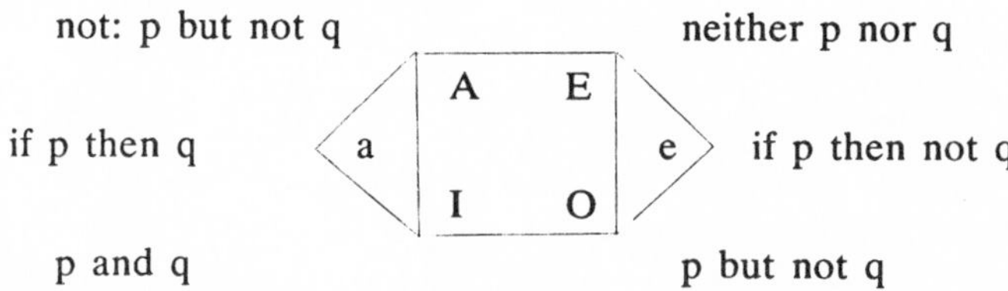

All the relations which hold for the categorical square also hold for this compound square. In particular, whenever I and O are both false a and e are undefined (A and I are then both true since LEM, as always, still holds). Vacuousity occurs whenever both I and O are false, i.e., whenever 'p' is false. In such cases both a and e are undefined. Just as universal forms are defined by particular quanity, denial, and predicate negation (e.g., 'every S is P' = 'no S isn't P' = 'not an S is nonP') only for nonvacuous cases, so conditional forms are defined by negation and conjunction (e.g., 'if p then q' = 'it is not the case that p but not q') only when 'p' is true. When 'p' is false we have vacuousity.

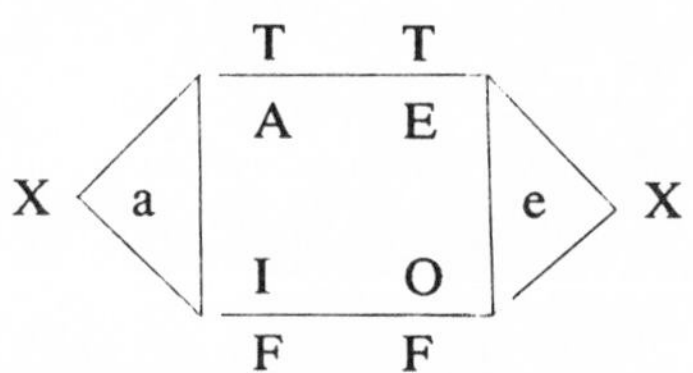

Given the usual notion of material implication now in use, a false sentence materially implies every sentence. For example, 'Napolean won at Waterloo' materially implies '2+2=6.' Indeed, this notion of material implication, along with the standard parsing of universals in terms of conditionals, also leads to the paradoxical view that every predicate holds of an empty subject. For example, since there are no unicorns, every unicorn is blue. For 'every unicorn is blue' is parsed as 'if anything is a unicorn then it is blue,' a conditional whose antecedent is false. But on our view, both 'if Napolean won at Waterloo then 2+2=6' and 'every unicorn is blue' are undefined since their corresponding I and O forms are all false. While in nonvacuous cases A=a and E=e, the paradoxes of material implication and existential import are easily avoided by the recognition of A/a and E/e distinctions for vacuous cases.

The augmented square of opposition preserves the relations found on the old square while at the same time recognizing linguistic and ontological distinctions which result in a richer array of logical relations. The added information displayed on an augmented square permits it to have a much wider range of applicability and far greater flexibility than has hitherto been suspected.

Chapter VI

Endnotes

* This essay first appeared in *Logique et Analyse*, 107 (1984).

1. See Englebretsen (1975b), (1976a), and (1984).

2. See note 4 of Chapter I above.

3. Sommers (1982).

4. For arguments or discussions see Vendler (1967), Chapter 2, Chastain (1975), Sommers (1978) and (1982) and Paduceva (1970).

5. For more on reference see Sommers (1975) and Englebretsen (1980c).

6. In addition to Sommers' logical studies see Englebretsen (1980), (1981) and (1981a).

7. See Sommers (1982), Appendix B and Englebretsen (1981).

8. Brentano (1956) defines A and E in a similar fashion. I and O are given as 'There is an S which is...' and A and E are then the contradictories of these, viz. 'There is not an S which is....'

9. Geach is so tempted. See his (1972), 2.5.

10. *On Interpretation*, 24b7-10; *Metaphysics*, 101b15-17.

11. See Sommers (1965a) and Englebretsen (1972), (1973a) and (1975b).

12. The above remarks are a drastic, but hopefully accurate summary of Sommers' arguments in Chapter 10 of (1982).

13. See the discussion in Englebretsen (1973a).

14. See especially Sommers (1959), (1963a) and (1971).

15. I'm ignoring here a subtle, but often very important, distinction between predicability and spanning. See Englebretsen (1971) and (1972) and Sayward and Voss (1972).

16. See Sommers (1959), (1963a) and (1965a) and Englebretsen (1969) and (1974a).

17. But see Englebretsen (1980a).

18. Leibniz (1966), p. 115.

19. Sommers (1969), (1970) and (1982).

20. See, for example, Leibniz (1966), pp. 17 and 66.

21. In Sommers (1982).

22. See Peirce (1933), 4.3.

Chapter VII

ON THE LOGIC OF PHRASAL CONJUNCTIONS*

Most contemporary logical theorists take reference to be the function of names (Geach) or quantifiable pronouns (Quine). In contrast, I want to argue for a more traditional theory. This theory holds that reference is the unique and proper province of quantified expressions. A quantified expression is a syntactical complex of a quantifier and a term. A term is a denoting word or phrase. The denotation of a term is its extension. Thus, 'if,' 'or,' 'whenever,' 'all,' etc., are not terms--they are formatives. Every term has a denotation. No formative has a denotation.[1] Nothing in logic requires that all of the denotation of a term exist. A quantifier is a nondenoting formative which, in concatenation with a term, determines a subset of the denotation of that term as the reference of the expression. Thus, while terms denote, quantified terms refer.[2] On this traditional view a universally quantified term refers to its entire denotation (makes distributive reference); a particularly quantified term refers to an undetermined (not necessarily undeterminable) subset (not necessarily proper) of its denotation (makes undistributive reference).

Contemporary theorists have largely abandoned the logic (syllogistic) which was based on the theory of logical syntax containing these insights about reference. They did so for a very large number of reasons. One of these reasons was that there were inferences which could not be adequately analyzed by the old theory. Inferences involving relationals and inferences involving identities are the most often cited examples. Yet the traditional theory of logical syntax can analyze such inferences. In his recent work F. Sommers has shown how to do this by extending the scope of the old theory in obvious ways.[3] The fact that the traditional theory is defensible against the attacks of contemporary logical theorists is itself a strong argument in favor of not abandoning the old theory too soon. What I want to argue is that the traditional theory has not only the unsuspected powers that Sommers has discovered (which make it comparable in scope to the standard first order predicate logic with identity), but that it has powers beyond those of the contemporary standard logic.

In particular, I want to argue that there are inferences (those involving phrasal conjunctions) which cannot be adequately analyzed by the standard theory, but which can be adequately analyzed by the traditional theory. Examples of sentences containing phrasal conjunctions are:

(1) Socrates and Plato are Greek
(2) Socrates and Plato weigh 300 pounds
(3) Aristotle admires Socrates and Plato
(4) The piano was carried upstairs by Socrates and Plato
(5) Plato and his teacher are Greek
(6) The teacher of Plato and the teacher of Aristotle are Greek

These sentences admit phrasal conjunctions (conjunctions of singular referring terms) into referential positions. Some theorists have argued that any such sentence is analyzable as a conjunction of subsentences.[4] This works well enough for some sentences, like (1), (3), (5) and (6). Thus:

(1.1) Socrates is Greek and Plato is Greek
(3.1) Aristotle admires Socrates and Aristotle admires Plato
(5.1) Plato is Greek and his teacher is Greek
(6.1) The teacher of Plato is Greek and the teacher of Aristotle is Greek

There are logically valid inferences from (1) to (1.1), from (3) to (3.1), from (5) to (5.1) and from (6) to (6.1). But there are no such inferences from (2) to (2.1) and from (4) to (4.1).

(2.1) Socrates weighs 300 pounds and Plato weighs 300 pounds

(4.1) The piano was carried upstairs by Socrates and the piano was carried upstairs by Plato

What is required for an adequate analysis of these kinds of inferences (both the valid and invalid ones) is a proper account of the logical form of sentences containing phrasal conjunctions. That the phrasal conjunctions in sentences like (2) and (4) (unlike those in (1), (3), (5) and (6)) are indissoluble ought to be reflected in the logical forms of those sentences. To put it simply, the contemporary theory of logical syntax embedded in the standard calculus is unable to reveal a difference in logical form between the phrasal conjunction in (1) and the phrasal conjunction in (2).[5] The traditional theory can do this.

To see how this is so, consider first how singular terms (such as names and definite descriptions) are incorporated into the traditional syntactical theory. On that account, as I said, reference is achieved only by quantified terms. So how can a singular term refer? Scholastic logicians generally held that singular terms are implicitly universal in quantity. This is because reference by a singular is always distributive. This is a simple way of incorporating singulars into the traditional theory of referential expressions. But it doesn't quite capture the syntactical uniqueness of singular terms. They seem to be more than just universally quantified terms with suppressed quantifiers. After all, 'men' in 'men are mortal' is such a quantified term, implicitly universal, yet it is not singular. Leibniz recognized that there are valid Inferences, especially third figure syllogisms, with singular conclusions. For example,

All men are rational.
Some man is Socrates.
Therefore, Socrates is rational.

The conclusion must be particular here. Such considerations led Leibniz to conclude that, while singulars are implicitly quantified, it doesn't matter which quantifier is chosen since they carry both implicitly.[6] This idea makes good sense in the light of the traditional theory. The reference of 'all Socrates' is the entire denotation of 'Socrates,' which is just Socrates. The reference of 'some Socrates' is a part of the denotation of 'Socrates,' which, having but one part, is, again, just Socrates. It doesn't matter whether reference to Socrates is made by universally or particularly quantifying 'Socrates.' The reference is the same in both cases. That explains why we dispense with any overt sign of quantity for singulars in natural language. This Leibnizian idea that singular terms have implicit "wild" quantity has been thoroughly exploited by Sommers in his attempts to extend the traditional theory.[7]

In general, if 'T' is a term with $a_1 \ldots a_n$ as its denotation, then there is (or we can always arbitrarily create) another term '$\lfloor a_1 \ldots a_n \rfloor$' with the same denotation. The converse need not hold. For example, I can replace '⌊Russell, Whitehead⌋ were British' with 'The authors of *Principia Mathematica* were British,' or '⌊3, 5, 7⌋ are odd' with 'every prime between 2 and 9 is odd'; but I have no comparable substitute for '⌊Kant, Frege⌋ were Prussian.' In other words, we sometimes want to refer to things for which we have no natural language-denoting term. It's as if we have the denotation but not the term. We are forced to make do then with what we have. English sentences whose logical forms are

every $\lfloor a_1 \ldots a_n \rfloor$ is F

are rendered naturally by the use of phrasal conjunctions. We can think of a term like $\lfloor a_1 \ldots a_n \rfloor$ as a denoting term like any other. Since it carries its denotation on its face, we might call it a term of "explicit denotation." Quantifing it results in a referring expression. When universally quantified it, like any universally quantified term, refers to its entire denotation. Thus, 'every $\lfloor a_1 \ldots a_n \rfloor$' refers distributively, i.e., to a_1 and...and a_n. The logical forms, then, of (1), (3), (5) and (6) are

(1.2) Every ⌊Socrates, Plato⌋ is Greek
(3.2) Aristotle admires every ⌊Socrates, Plato⌋
(5.2) Every ⌊Plato, his teacher⌋ is Greek

(6.2) Every ⌊the teacher of Plato, the teacher of Aristotle⌋ is Greek

It should be noted that the particular quantification of a term of explicit denotation is a phrasal disjunction. Such a term, when so quantified, like any particularly quantified term, makes undistributive reference. So a sentence like 'some ⌊Socrates Plato⌋ taught Aristotle' is rendered naturally in English with a phrasal disjunction: 'Socrates or Plato taught Aristotle.' It might also be noted that since, for any singular terms 's,' 's'='⌊s⌋,' we have another argument for the wild quantity of singulars, i.e., 'every s'='some s,' since 'every ⌊s⌋'='s and...and s' and 'some ⌊s⌋'='s or...or s,' and 's and...and s'='s or...or s' = 's.'

Treating phrasal conjunctions like those in (1), (3), (5) and (6) as universally quantified terms of the form '$\lfloor a_1 \ldots a_n \rfloor$' allows us to incorporate them into the traditional theory of referring expressions, and permits an easy account of the validity of such inferences as (1), therefore (1.3).

(1.3) Socrates is Greek

Such inferences are enthemematic. G. Massey has justifiably cautioned against what he terms "the enthemematic ploy."[8] The ploy consists of taking the corresponding conditional of an inference (which is taken to be intuitively but not fully formally valid) as a suppressed premise. Since such a manoeuver could be used to render any inference valid it ought to be avoided on all occasions. But this is no condemnation of enthememes in general. Inferences like 'John is a man. So John is rational' are valid enthememes. The suppressed premise is analytic and is not the corresponding conditional.[9] Sentences of the following form are always analytic.

(7) every $\lfloor a_1 \ldots a_n \rfloor$ is $\lfloor a_1 \ldots a_n, a_m \ldots \rfloor$

Using them as the suppressed premises of enthememes will be logically harmless. Consider now the inference from (1) to (1.3). The entire enthememe has the form:

every ⌊Socrates, Plato⌋ is Greek
every ⌊Socrates⌋ is ⌊Socrates, Plato⌋
therefore, every ⌊Socrates⌋ is Greek

a Barbara syllogism.

The inference from (2) to (2.3)

(2.3) Socrates weighs 300 pounds

is invalid. If two inferences have the same natural language surface form, yet one is formally valid and the other invalid, they must differ in logical form. In particular, we need to look for a difference in logical form between the phrasal conjunction 'Socrates and Plato' as it appears in (1) and as it appears in (2). (2) cannot be formulated as 'every ⌊Socrates, Plato⌋ weighs 300 pounds.' For we are clearly not, in normal situations, making a distributive reference to both Socrates and Plato in (2).

Just as we can arbitrarily form a term '$\lfloor a_1 \ldots a_n \rfloor$' with $a_1 \ldots a_n$ as its denotation, we can form a term '$)a_1 \ldots a_n($' which is a singular term, and having as its denotation *the group* consisting of $a_1 \ldots a_n$. The terms $\lfloor a_1 \ldots a_n \rfloor$ and '$)a_1 \ldots a_n($' differ in important ways. While the former is sometimes singular (as when n=1), the latter is always singular. While (7) is analytic, (8) is not.

(8) $)a_1 \ldots a_n($ is $)a_1 \ldots a_n, a_m \ldots($

Also, every $\lfloor a_1 \ldots a_n \rfloor$ is F if and only if a_1 is F and...and a_n is F; but it is not the case that $)a_1 \ldots a_n($ is F if and only if a_1 is F and...and a_n is F. However, for singular 's,' 's'=')s(.' Indeed, for singular 's,' 's'='⌊s⌋'=')s(.' We can think of '⌊...⌋' as a function on one or more singular terms which forms a general term, and ')..(' as a function on one or more singular terms which forms a new singular term.

By treating some phrasal conjunctions (e.g., those in (2) and (4)) as group-denoting terms, which are always singular (having wild quantity), phrasal conjunctions are fully incorporated into the traditional account. The inference from (2) to (2.3) is seen to be invalid since its formal validity (as an enthememe) would require a suppressed premise which does not hold. Thus:

)Socrates, Plato(weigh 300 pounds
)Socrates(is)Socrates, Plato(
therefore,)Socrates weighs(300 pounds

Phrasal conjunctions are either group terms of the form '$)a_1 \ldots a_n($' or terms of explicit denotation of the form '$\lfloor a_1 \ldots a_n \rfloor$,' Each such term, when used as a referring expression, is implicitly quantified. For the latter kind of term, conjunction ('and' in English) is the sign of universal quantity, disjunction ('or') the sign of particular quantity. For group terms the implicit quantity is wild since they are always singular. In any case, the presence of singular terms and of phrasal conjunction of singular terms presents no serious challenge to the traditional notion that all reference is achieved by (possibly implicitly) quantified terms. In this respect the old theory has a scope of application exceeding that of the new one.[10] Modern logical theorists have not been able to adequately account for the shaded logic of phrasal conjunction because they have preferred, like the drunk searching for his keys under the lamp post, to stay in the brighter light of names and pronouns.

Chapter VII

Endnotes

* A Rumanian version of this chapter first appeared as Englebretsen (1984g).

1. For a fuller account of this distinction see Sommers (1973a).

2. See Sommers (1975) and Englebretsen (1980c).

3. In addition to his papers cited in notes 1 and 2 above see the works cited in notes 4 and 5 of Chapter I above.

4. See Gleitman (1965).

5. See Massey (1976).

6. Leibniz's notion is found in "A Paper on 'Some Logical Difficulties,'" in Leibniz (1966).

7. In addition to Sommers' work see Noah (1973), Wald (1979), and Englebretsen (1980a) and (1982).

8. Massey (1976).

9. Innocuous enthememes are further characterized in Englebretsen (1979).

10. A good example of the modern theorist's fruitless struggle with phrasal conjunctions is found in Strawson (1970).

Chapter VIII

COMPOUND TERMS*

Simplify, simplify.
Thoreau

General terms can appear in both subjects and predicates of sentences. Thus, in 'Some singer is a performer' and 'Every performer is vain' the general term singer occurs in the subject of the first sentence, the term 'vain' appears in the predicate of the second, and the term 'performer' occurs in the predicate of the first and in the subject of the second. It is this ability of general terms to occupy both subject and predicate positions which makes syllogistic possible. General terms can be quantified (to form subjects); or they can be qualified (to form predicates). In our examples above 'performer' was qualified in the first sentence and quantified in the second.

General terms can be negated. Thus, in 'Some nonsingers are performers' and 'No one who is kind is unloved' the terms 'singers' and 'loved' are negated. It is the ability to form term negations which makes conversion, obversion and contraposition possible.

General terms can be conjoined or disjoined to form compound terms. Thus, the terms 'singer' and 'dancer' are compounded in 'Every star is a singer and dancer' and 'Some singer or dancer will play the lead.'

The ideas that general terms could be quantified, qualified, negated, conjoined and disjoined were generally understood by traditional, preFregean logicians and grammarians. None of these ideas however has survived as part of the standard modern predicate logic. According to the theory of logical syntax upon which modern predicate logic is based, there are two kinds of sentences--those with no syntactical complexity at all (atomic sentences), and those with some syntactical complexity (molecular sentences). Atomic sentences always consist of a predicate and one or more subjects. Predicates are always general terms; subjects are always singular terms (names, pronouns). Atomic sentences never contain any formative expressions. In particular, they never contain quantifiers, qualifiers, negators, conjoiners or disjoiners. The presence of any formative element is a guaranteed indication of syntactical complexity (molecularity). Modern predicate logic gives no logical role at all to qualifiers (copulae). The other formatives, moreover, are never allowed to operate on terms (even general terms). Quantification, negation, conjunction, disjunction and any other "truth-function" can only apply to sentences to form syntactically more complex sentences. Thus 'John is a singer' is an atomic sentence consisting of two syntactically simple expressions: a predicate 'is a singer' (symbolized 'S') and a subject 'John' (symbolized 'j'). The sentence has the form 'Sj.' Sentences such as 'John is not a singer,' 'John is a singer and a dancer' and 'Some singer is a dancer' are all molecular, constructed from atomic sentences and formatives on them. 'John is not a singer' is taken as the negation of 'John is a singer' ('~(Sj)'). 'John is a singer and a dancer' is viewed as a conjunction of the two atomic sentences 'John is a singer' and 'John is a dancer' ('Sj & Dj'). 'Some singer is a dancer' is seen as a quantifier applied to a molecular sentence. The quantifier is 'there exists some thing such that' and the molecular sentences is a conjunction of 'it is a singer' and 'it is a dancer' (symbolically: '(∃x)(Sx & Dx)'). Here is a clear example of how the modern predicate logician's identification of predicates with general terms forces him or her into creating pronominal subjects. The traditional logician simply took 'Some singer is a performer' to be a quantified term concatenated with a qualified term. The traditional logician saw this sentence as saying of some singer that it is a dancer. The contemporary logician sees it as saying of some thing that it is a singer and it is a dancer.[1]

In modern predicate logic all logical formatives are sentential. (This is why the sentential calculus is said to be primary and foundational for the predicate calculus.) Natural language sentences containing, for example, term negations or conjunctive terms must first be paraphrased as sentences containing sentential negation and sentential conjunction. Traditional logic was a logic of terms. In the late nineteenth century traditional logic was abandoned in favor of modern predicate logic because the new logic could do so much more than the old. Unfortunately, a hidden assumption in this process was that traditional logic (the old syllogistic) exhausted the capacities of term logic. Nonetheless, the assumption was unfounded--indeed, it is false. It is possible to construct a term logic which, while preserving many of the insights of traditional logic, goes far beyond it in terms of its powers of expression and inference. In fact, such a logic (i.e., a new syllogistic) enjoys certain important advantages over modern predicate logic: its logical syntax is more natural (in that it formulates sentences in ways much closer to their grammatical forms); it is simpler (in that the number of principles required to account for valid inference is relatively small, and those principles are easily stated); and it is more powerful (in that there are inferences, often quite simple ones, for which it, but not the standard logic, can account).[2]

For the term logician all formatives operate on terms. Any term can be quantified, qualified, negated, or conjoined or disjoined with another term. Let us call terms which are negated or which are conjunctions or disjunctions of terms compound terms. Since compound terms are terms they can be quantified, qualified, negated, or conjoined or disjoined with other terms. General terms, like 'singer,' 'dancer,' 'vain,' and the like are paradigmatic for terms. But singular terms (viz. names and pronouns) are terms as well. So are entire sentences (called sentential terms). And just as general terms can be quantified, qualified, negated, etc., so too, *mutatis mutandis*, can singular and sentential terms. The contemporary standard logician recognizes a bit of this of course. He or she allows the negation, conjunction and disjunction of sentential terms, and the quantification (in some sense) of pronouns--but that's it.

For the term logician the distinctions between general terms, singular terms, and sentential terms are semantic only. Syntactically they are all on all fours with one another. Thus, the logic of singulars is no different from the logic of generals. And a logic which recognizes this enjoys the advantage of predicating (qualifying) singulars, thus eliminating the need to have a special "identity theory" appended to the main system.[3] Moreover, a logic which recognizes sentences as terms, fit for all the logical roles reserved for any term, enjoys the not insubstantial advantage of having a single common logic of terms and sentences. So a logic of sentences is not primitive and basic to a logic of terms. It is a part of the logic of terms. Such a logic gets by with one calculus instead of two. Remember Thoreau.

Most contemporary philosophers and logicians find many of the ideas presented above hard to swallow. Yet each can be, and has been, well-defended. In what follows I want to explore just one of the ideas central to a logic of terms. I want to look at compound singular terms. This is actually two ideas: that singular terms can be negated, and that singular terms can be conjoined or disjoined. Since my thoughts on negated singulars have been fairly widely broadcast,[4] I will merely say that the negation of a singular is not a singular. The mistaken belief that the negation of a singular term is a singular term (moreover, one which, if it denoted, could only denote an impossible object) has been the source of much confusion for both standard and term logicians as well as many philosophers and grammarians.

So I shall confine myself here to conjunctions and disjunctions of singular terms. In particular, I want to examine the logical syntax of such compound terms.[5] In a natural language such as English we conjoin and disjoin terms by use of such formative expressions as ‘and,’ ‘as well as,’ ‘also,’ ‘but’ and the like, and ‘or.’ Let ‘and’ be our standard formative for conjunction, and let ‘or’ be our standard formative for disjunction. Now it is wise to remember that not every use of ‘and’ is aimed toward conjunction. Consider the sentence ‘Some logicians are scholars and gentlemen.’ Here the compound term ‘scholars and gentlemen’ is the result of conjoining the two terms ‘scholar’ and ‘gentlemen.’ We could paraphrase the compound term as ‘both scholars and gentlemen.’ Whenever a compound predicate-term of the form ‘P and Q’ can be paraphrased as ‘both P and Q’ we will call it a *genuine* conjunction. Not all conjunctions are genuine. Consider the sentence ‘All the guests were men and women’ (presumably, no children were invited). Here the compound term ‘men and women’ could not be paraphrased as ‘both men and women.’ The conjunction here is a *pseudo* conjunction. Pseudo conjunctions are, logically, disjunctions. We could paraphrase our sentence as ‘All the guests were men or women.’

I am going to symbolize a genuine conjunction, e.g., ‘P and Q’ (= ‘both P and Q’) as ‘+P+Q.’[6] We can read the first + as ‘both’ and the second as ‘and.’ Compound terms, at least conjunctions and disjunctions, will always be placed between angular brackets; thus, ‘<+P+Q>.’ Disjunctions are defined in terms of conjunction and negation. A compound like ‘P or Q’ is first paraphrased as ‘not both nonP and nonQ.’ Using the minus sign, –, for negation, this is formulated as ‘–<+(–P)+(–Q)>.’ Distributing the minus sign gives us, finally, ‘<–(–P)–(–Q)>,’ or more simply, ‘<– –P– –Q>.’ We could replace double negations here with pluses, but we will resist the temptation to do so, keeping the formulations for ‘P and Q’ and ‘P or Q’ distinct. We can read our first pair of –’s as ‘either’ and our second pair as ‘or.’ Since ‘or,’ unlike ‘and,’ is a defined formative, we must always be prepared to reformulate any disjunction in terms of negations and conjunction.

It is time, at last, to turn to singulars. Consider the quite unexceptional English sentences ‘Al and Betty are together’ and ‘Al and Betty are happy.' How would these be formulated in the canonical notation of standard predicate logic? Usually like this: 'Wab' and 'Ha & Hb' (here 'W' is read as 'is with' and 'H' is read as 'is happy'). But now why is 'Al and Betty are together' taken as an atomic sentence with a binary predicate while 'Al and Betty are happy' is taken as a molecular sentence conjoining the two atomic sentences 'Al is happy' and 'Betty is happy'? Surface appearances suggest, contrarily, that 'Al and Betty are together' and 'Al and Betty are happy' should have the same logical structure. They certainly have the same grammatical structure. (Of course, the standard predicate logician has never tired of denigrating natural language forms, denying that natural language even has a logic.) The term logician can preserve the grammatical similarity between 'Al and Betty are together' and 'Al and Betty are happy' in their logical forms. First, notice that each conjunction here is pseudo (remember the 'men and women' example). In both sentences 'Al and Betty' is paraphrased in the same way and formulated as '<– –a– –b>' (we use lowercase letters for singulars just to remind ourselves of their singularity). We could read this expression back into English as 'things which are either Al or Betty.' Now since 'Al and Betty' (read as a pseudo conjunction, i.e., a disjunction) is a subject term here, it is, implicitly, quantified. In this case the logical quantity is universal (compare: 'Apples and oranges are fruits'). The two sentences have the forms '(every) <– –a– –b> is T' and '(every) <– –a– –b> is H' (where 'T' is read 'together'; we will refrain from symbolizing quantifiers and qualifiers here).

Consider next a straightforward disjunction such as 'either Al or Betty' (as in 'The winner is either Al or Betty' and 'Either Al or Betty is a winner'). Here the compound term 'either Al or Betty' is simply transcribed as '<– –a– –b>.' The first example would have the form "The winner is '<– –a– –b>,'" where the compound term is predicated (in this case, affirmed) of the subject. In the second example the compound is the subject term and therefore is, implicitly, logically quantified. Here the understood quantity is clearly particular. Thus: '(some) <– –a– –b> is W.'

Suppose now that we have a sentence which we formulate as '(every) <– –a– –b> is T.' How might that sentence look in English? Perhaps 'Both Al and Betty taught school.' And now, I think, we can begin to see the secret of singular compounds. Conjunctions of singulars, terms of the form 'a and b' are pseudo conjunctions. This is so because, assuming the uniqueness of each individual, a singular only applies to (denotes) one individual. A thing can be both short and fat, both red and square, or both even and prime, but nothing can be both Al and Betty, Paris and London, or 2 and the square root of 11. When it comes to compounds of singulars, such terms are either genuine disjunction or pseudo conjunctions--none is genuine conjunction. How then do we distinguish between these two kinds of disjunctions? Well, as it turns out, when such terms are predicated (qualified) there is no difference--they are always simply disjunctions. Sentences formed as 'Some T is <– –a– –b>' and 'Every T is <– –a– –b>' would be read respectively as, e.g., 'Some teacher is either Al or Betty' and 'Every teacher is Al or Betty.' Conversion of each such sentence would result in a sentence of the general form 'Some
<– –a– –b> is T.' And it's the quantity of quantified compounds of singulars which determines whether they are genuine disjunctions or pseudo conjunctions. When the logical quantity is particular the compound is a genuine disjunction. When the logical quantity is universal the compound is a pseudo conjunction. Thus 'Some <– –a– –b> is T' and 'Every <– –a– –b> is T' could be read as 'Al or Betty teach' and 'Al and Betty teach' respectively.

Some philosophers have had some confused and muddled things to say about compounds of singulars.[7] This has usually been due to their failure to appreciate two important points. First, they take all pseudo conjunctions of singulars to be genuine conjunctions. This has lead them to say of subject terms such as 'Al and Betty' that they must name, denote, impossible objects, since what 'Al and Betty' denotes must have the properties of both Al and Betty; but many of the pairs of properties which Al and Betty have are logically exclusive (e.g., male/female, tall/short, bald/blond, etc.). The corrective here is simply the recognition that no conjunction of singulars is genuine. Their second misunderstanding is more basic. The standard predicate logician has taught them that all logical subjects must be singular. It follows from this that sentences like 'Al and Betty teach' and 'Al or Betty teach' either have singular subjects or are compounds of sentences with singular subjects. But, since treating compounds such as 'Al and Betty' and 'Al or Betty' as names of single individuals is nonsense, the functional expressions here must be seen as sentential, contrary to surface appearances. The simple remedy in this case is to follow the term logician in giving up the dogma that all logical subjects must be singular.[8]

I have said that no conjunction of singulars is genuine. But not all of them are pseudo conjunctions (disjunctions). Sometimes terms like 'Al and Betty' really are singulars, denoting a single object. But the object denoted is certainly not, *per impossible*, some strange beast enjoying all the properties of both Al and Betty. When we say 'Al and Betty are together,' 'Al and Betty are happy,' 'Al and Betty teach' we make reference to everyone who is either Al or Betty (these are pseudo conjunctions with implicit universal quantity). But consider now sentences such as 'Al and Betty carried the piano upstairs' and 'Al and Betty beat Carol and Don at bridge.' When 'Al and Betty' is the subject term of a sentence, and a pseudo conjunction (thus having implicit universal quantity), what is true of the conjunction is true of each member of that conjunct. If Al and Betty are happy then Al is happy and Betty is happy. But if Al and Betty carried the piano upstairs it need not follow that Al carried the piano upstairs and that Betty carried the piano upstairs. In such cases we could not formulate the sentence using '(every) <– –a– –b>.' For in these cases what is being denoted by 'Al and Betty' is a single individual. This individual is not a person; it is not Al; it is not Betty; it is not any impossible object sharing all of both Al and Betty's properties. I will call the individual denoted by such compounds a *team*. What is true of a team need not be true of each (or even any) of its members. For example,

Al did not beat Carol and Don at bridge--the team consisting of him and Betty did. Gary Carter did not win the 1986 World Series--The Mets won the series. His satisfaction (and pay) came not from winning the series, but from being a member of the team that won the series. Where 'Al and Betty are happy' has the form '(every) <– –a– –b> is H,' 'Al and Betty won the bridge game' has a form something like '(the) [a,b] is W' (where '[a,b,c...n]' denotes a team whose members are a, b, c...n). (Another example of a team name is 'Nicolas Bourbaki.')

Sometimes proper names can be used to replace conjunctions of singulars. Instead of saying that Al and Betty are teachers, we might say 'The Smiths are teachers.' In such a case 'Smiths' is a general term, not a singular. We could paraphrase the sentence as 'All of the Smiths are teachers,' and formulate it as 'Every S is T.' Here the general term 'Smiths' replaces the general term 'Al and Betty' (i.e., 'S'='<– –a– –b>'). Of course, when Al and Betty are not related and sharing the same family name we do not, normally, have such an equivalence.

Often proper names can be used to replace team-denoting conjunctions. Indeed, these are just team names. If Al and Betty Smith beat Carol and Don Green at bridge, I might very well say 'The Smiths beat the Greens at bridge,' where 'Smiths' and 'Greens' are not general terms replacing pseudo conjunctions, but singular terms denoting teams (thus replacing conjunctive team names). I could say 'Gary, Ray, Mookie, Dwight,...won the series.' But I would probably say 'The Mets won the series.'

If a term logic can be constructed which is natural, simple and effective it deserves some attention from logicians and philosophers (and probably linguists). I have tried to show here that, when it comes to compound singulars, such a logic can provide a natural, simple and effective analysis. Compound general terms, compound sentences, relationals, and more can be handled just as naturally, simply and effectively by a logic of terms.[9]

Chapter VIII

Endnotes

* Discussions with Fred Sommers and Bill Shearson (that's a pseudo conjunction!) helped me get clearer about this topic. Thomas Coats and James Blevins made several valuable suggestions as well.

1. For more on syntactical complexity see Englebretsen (1984e).

2. See especially Sommers, forthcoming.

3. See Sommers (1969) and Chapter 6 of (1982) and Englebretsen (1982).

4. See Englebretsen (1985a), (1985b), (1985d) and (1987c).

5. For a different, but compatible, theory see Englebretsen (1984g), (1985a) and (1987c).

6. The symbolization follows Sommers.

7. The confusion is found in Strawson (1970) and (1974), Zemach (1981), Linsky and King-Farlow (1984) and Geach (1972a). For brief responses to Strawson, Zemach and Geach see the items cited in note 4 above.

8. The dogma is explored in detail by Sommers in (1967). Note that while the traditional logician chose to admit only one logical subject for each sentence (but permitted logical subjects to have any number of referents), the contemporary logician chooses to admit only one referent for each logical subject, i.e., all logical subjects are singular, (but permits sentences to have any number of logical subjects).

9. For a different approach to the kinds of compound terms discussed here see Link (1983) and (1984). See also Gleitman (1965) and Massey (1976).

Chapter IX

PRELIMINARY NOTES ON A NEW MODAL SYLLOGISTIC*

This essay consists of five parts. In section one we introduce the topic of modal syllogistic by examining the case of the two Barbaras found in *Prior Analytics*. In the second section we briefly review certain aspects of the "new syllogistic" developed in recent years by Fred Sommers. The next two sections examine some of the syntactic and semantic features of modal sentences *de dicto* and *de re* respectively. Our final section presents a preliminary sketch of what a syllogistic admitting both *de dicto* and *de re* modality would look like.

1. The following dilemma for modal syllogistic was once posed by Martha Kneale:

> If modal words modify predicates, there is no need for a special theory of *modal* syllogisms. For these are only ordinary assertoric syllogisms of which the premises have peculiar predicates. On the other hand, if modal words modify the whole statements to which they are attached, there is no need for a special modal *syllogistic*, since the rules determining the logical relations between modal statements are independent of the character of the propositions governed by the modal words (Kneale (1962), p. 91).

Syllogistic's author, Aristotle, was clearly aware of the *de dicto/de re* distinction alluded to here. In *De enterpretatione* he was at some pains to spell out various laws of modal propositional logic, where modal words are purely *de dicto*. But in *Prior Analytics* he attempted a modal syllogistic. The degree of his success in this is still an issue of debate. The question has never been whether or not a modal syllogistic could be formulated. Rather it has been whether one could be formulated which preserves both formal adequacy and what McCall, (1963), called "Aristotelicity." How important it is to preserve *all* of Aristotle's claims, insights, intuitions and hunches is itself debatable. There are places in Aristotle's work which are simply muddled. For example, despite the fact that he clearly saw the formal distinction between *de dicto* and *de re* sentences he sometimes takes corresponding pairs of de *dicto/de re* sentences as interchangeable. Thus, he states in *Prior Analytics* (25a29) the apodeictic negative universal as 'A necessarily applies to no B' and (30b12) as 'A possibly applies to no B.' Now the first is easily reformulated as

(1) Necessarily no B is A

The second would appear to be equivalent at first sight to

(2) Possibly no B is A

But taking (1) and (2) as equivalent is too gross an error to attribute to Aristotle, He surely took the second sentence to be a *de re* version of the first. That is

(3) No B is possibly A

Nonetheless, contra McCall, it was an error (though a more subtle one) to equate (1) and (3). For let 'A' be uniformly replaced by 'nonB.' We would then have the following substitutions.

(1.1) Necessarily no B is nonB
(3.1) No B is possibly nonB

Now (1.1) is obviously true ('No B is nonB' is analytic). But (3.1) is just false. Consider: 'No bachelor is possibly a nonbachelor (='No bachelor is a possible nonbachelor'). Surely Michael Jackson is an actual bachelor who is a possible nonbachelor; it's only a contingent matter that he is a bachelor.

If (1) and (3) are nonequivalent the semantic as well as formal *de dicto/de re* distinction must be preserved. And, while Aristotle seems to have blurred the difference here, he must have been recognizing it clearly (if not explicitly) elsewhere. Aristotle's discussion of "the two Barbaras" (*Prior Analytics* 30a21-28) is a good place to see this. Aristotle is talking there about syllogisms with mixed categorical and apodeictic premises and an apodeictic conclusion. Some are valid--others are not. For example, Barbara with the major and conclusion apodeictic is valid, but Barbara with the minor and conclusion apodeictic is invalid. Let us call these "Barbara I" and "Barbara II" respectively.[1] They are:

Barbara I: A necessarily applies to all B
B applies to all C
So A necessarily applies to all C

Barbara II: A applies to all B
B necessarily applies to all C
So A necessarily applies to all C

If we interpret the modal words in these as explicitly *de dicto* we get

Barbara I(d): Necessarily all B is A
All C is B
So necessarily all C is A

Barbara II(d): All B is A
Necessarily all C is B
So necessarily all C is A

Now Aristotle's thesis is that Barbara I, but not Barbara II, is valid. Does this difference hold for the *de dicto* readings? Barbara II(d) *is* invalid. A counter-argument would be to substitute C for B, Thus:

All C is A
Necessarily all C is C
So necessarily all C is A

Here the assertoric/modal (viz. apodeictic) distinction guarantees its invalidity. But a similar counter-argument can just as well be used to show the invalidity of Barbara I(d).[2] Let A be a substitute for B, then

Necessarily all A is A
All C is A
So necessarily all C is A

So let us give the two Barbaras an explicitly *de re* reading.

Barbara I(r): All B is necessarily A
All C is B
So all C is necessarily A

Barbara II(r): All B is A
All C is necessarily B
So all C is necessarily A

There is no question that Barbara II(r) is formally invalid. But what of Barbara I(r)? Suppose we try to show it invalid by a counter-argument parallel to the one used against Barbara I(d). Substituting A for B we get

All A is necessarily A
All C is A
So all C is necessarily A

Unlike the major of the (d) counter-argument, this major is not analytic. There is at least one false instance of it. Consider: 'All bachelors are necessarily bachelors.' Again, Michael Jackson is a bachelor who is a possible nonbachelor. So calling the counter-argument valid would not force us to derive the conclusion from the minor alone (in turn collapsing the assertoric/apodeictic distinction). So this is not a counter-argument to Barbara I(r). In fact, Barbara I(r) is valid. When *de re* modal expressions are seen as forming new "peculiar" predicates (i.e., predicate-terms), i.e., 'All B is (necessarily A)' rather than 'All B (is necessarily) A,' the formal validity of Barbara I(r) is guaranteed by the formal validity of Barbara. Thus it is only the *de re* readings which preserve Aristotle's thesis that Barbara I, but not Barbara II is valid.

2. A syllogistic of *de re* modality can be easily constructed. It is simply a fragment of the assertoric syllogistic with modalized terms. But both syllogistic and *de re* modality have lost favor in this century. While it has had its recent defenders, *de re* modality has been denigrated by logicians like Quine who are repelled by what are perceived as its essentialist consequences. And syllogistic is even more broadly scorned. At worst it is seen as downright mistaken. At best it is seen as merely a fragment of the standard first order function calculus.

Quine's worries are caused by the possibility of allowing modal expressions to function within the scope of a quantifier. Such "quantifying in" through modal contexts cannot be permitted since such contexts are referentially opaque. But these worries vanish if both quantifiers and modals are taken as term functors rather than sentence functors. Of course the real trick here is construing quantifiers as term functors. To do so would require an effective logic of terms. Ironically, Quine himself has constructed one version of such a logic--the predicate functor algebra (see Noah, (1980), and Quine, (1966), (1976), (1976a), (1980), (1981a)). Another recent version of a term logic is Fred Sommers' "new syllogistic."

Sommers' logic deals with the canonical fragment of a natural language whose sentences either have a finitely and recursively specifiable syntax, or can be paraphrased as sentences which do. The canonical sentences are logically categorical, consisting of a subject and a predicate. A subject is a syntactically complex expression consisting of a quantifier and a term. A predicate is a syntactically complex expression consisting of a qualifier (copula) and a term. Terms are either syntactically simple or complex. A complex term is syntactically categorical. On this theory quantifiers are not functors operating on entire sentences. Instead they are functions on terms, yielding subjects. Quantifiers are universal or particular. Qualifiers are affirmative or negative. And terms come in positive/negative pairs. For example, 'red' and 'nonred' are such a pair.

The elementary sentences of Sommers' syllogistic are particular affirmations and their corresponding negatives (viz. contradictories). Letting S and P be any terms, the following represent elementary sentences.

1. An S is P
2. An S is nonP
3. Not an S is P (No S is P)
4. Not an S is nonP (No S is nonP)

Part of Sommers logic is a symbolic algorithm which achieves maximum efficacy with a minimum of symbolic devices. All particular quantifiers are reformulated by +'s. Affirmations are indicated by a +

preceding the formula. Negations are likewise indicated by –'s, which are also used to indicate negative terms. Positive terms are marked +. The sentences above are then symbolized as follows (where parentheses are, as usual, used to resolve questions of scope).

1.1 +(+(+S)+P)
2.1 +(+(+S)–P)
3.1 –(+(+S)+P)
4.1 –(+(+S)–P)

As in algebra, we can safely simplify these by suppressing initial +'s as well as the signs on positive subject terms. Thus:

1.2 +S+P
2.2 +S–P
3.2 –(+S+P)
4.2 –(+S–P)

Notice that now the first sign in 1.2 and 2.2 is a sign of quantity not quality (since these have no explicit quality sign they are understood as affirmations). Quantity signs are never suppressed.

Now, if we take the universal quantifier to be defined in terms of the particular[3] and symbolize it accordingly (by –), we have

1.3 +S+P
2.3 +S–P
3.3 –S–P
4.3 –S+P

Here all of the initial signs are signs of quantity and all sentences are taken affirmatively. Sentences 3.3 and 4.3 are achieved simply by distributing the quality signs (the initial –'s) in 3.2 and 4.2. We now read 4.3, for example, as 'Every S is P.'

Singular sentences, those with singular terms as subjects, are taken to be implicitly particular in quantity. They differ from other particulars only in that, given our extralogical knowledge of their singularity, we can infer from them their corresponding universals.[4]

It is easy to imagine how we might add to Sommers' system a "new modal syllogistic." At least, we can formulate modal sentences and specify, where necessary, additional rules of inference. We could add to our lexicon a modalized version of each term *per se*. Thus, if 'P' is a term then '◇P' (read 'possibly P' or 'a possible P') is a term.

Such modalized terms can themselves be negative as well as positive, so that we might have expressions of the form '–◇P,' etc. Finally, we could permit terms of any sort (e.g., positive or negative) to be modified by modality, resulting, for example, in such expressions as '◇–P' and '–◇–P.' By doing so we can take one sign of modality, say 'possibly' (◇) as primitive and define necessity (□) accordingly.

Inferences in such a system would be governed by the laws of immediate inference and the syllogism, augmented by laws such as

If +S+□P then +S+P
If –S+□P then –S+P
If +S+P then +S+◇P
If –S+P then –S+◇P

These brief remarks should be sufficient to show that, given the viability of Sommers' term logic, a formally adequate modal syllogistic could be devised.

3. Statements (statement-making sentences) are always made relative to some specifiable *domain* of discourse (what Sommers sometimes calls the statement's *amplitude*). Domains are coherent totalities specified by their mutually compatible constituents. Sets, whose members are always fixed, and worlds, whose memberships may vary, are both kinds of domains. (See Englebretsen, (1985f) and Sommers, (1982), (1987)). We say, 'A black horse won the Kentucky Derby in 1948,' and our statement is made relative to the actual world. It is the way the actual world is which determines the truth or falsity of our claim. But we also say, 'A winged horse was captured by Belerophon,' in which case our domain is clearly not the actual world, but rather the world of Greek mythology. I can assert 'There are winged horses' and also 'There are no winged horses' without contradiction as long as my two statements are not both made relative to a common domain. A logical relation between two statements requires that both can be made relative to a common domain.

On Sommers' theory, to be is to be a constituent of the domain in question. Any canonical statement can be read as an implicit claim that some kind of thing exists or does not exist in some given domain.

To say, with respect to the actual world, that some horse is black is implicitly to claim that horses are black horses in the actual world. To say, with respect to the mythological world, that some horses are winged is implicitly to claim that there are winged horses in the world of mythology. To say, with respect to the actual world, that no horses are winged is to claim implicitly that there are no winged horses in the actual world. And to say, with respect to the actual world, that every horse is tame is to claim implicitly that there are no wild (=nontame) horses in the actual world. Such implicit claims specify, in effect, the truth conditions for these statements. In general, every statement has both a *denotation* and a *signification* (see Chapter V). What a statement denotes is the domain relative to which it is used. What a statement signifies is a property of that domain--in particular, what Sommers calls a *constitutive* property. To say of anything that it has the constitutive property of being P-ish is to say that it has a P-thing (some thing which is P) as a constituent. A soup which has salt in it has the constitutive property of being salty (saltish). A domain which has a dog in it has the constitutive property of being dogish. More generally, to use 'An S is P' relative to domain d to make a statement is to characterize d as having the property of being SP-ish (written '[SP]'), i.e., as having as one of its constituents an S which is P. The implicit truth claim of any statement is that the domain which it denotes (viz. the one relative to which it is used) has the (constitutive) property it signifies.

All canonical statements are made relative to a determinate domain. But some noncanonical statements are made with respect to no determinate domain. These statements are made, we might say, relative to the domain of domains. To this group belong all statements of *de dicto* modality. For example, the statement 'Possibly some S are P' does not claim that S's which are P (SP-things) are in the actual world. Nor does it claim that SP-things are possibly in the actual world. It is a statement made with respect not just to the actual world, or any other particular domain, but rather with respect to all domains. (Sommers calls this "unrestricted amplitude.") It claim implicitly that SP-things exist in some (undetermined) domain. Used to make a statement, 'Necessarily some S are P' claims that SP-things are in every domain.

Let ∪d represent the union of all domains, and ∩d represent the intersection of all domains. Since *de dicto* apodeictic statements make a truth claim about every domain we will say that they are implicitly made relative to ∩d. Since *de dicto* problematic statements make a truth claim about some unspecified domain(s) we will say that they are implicitly made relative to ∪d. Letting [p] be the constitutive property signified by 'p' when used to make a statement, we can say that '□p' ('necessarily p') implicitly makes the truth claim that every domain is [p], i.e., ∩d is [p]. Also '◇p' ('possibly p') implicitly makes the truth claim that some domain is [p] , i.e., ∪d is [p].

Suppose I state that it is necessary that not p (by using '□–p'). My implicit truth claim is that every domain lacks the property [p]; in other words, no domain is (p), ∩d is un[p]. We display below the eight *de dicto* modalized categoricals along with their truth claims ('P' means 'nonP').

Table 1.		Formula	Truth claim
	□ a :	□(–S+P)	∩d is un[S$\bar{P}$]
	□ e :	□(–S–P)	∩d is un[SP]
	□ i :	□(+S+P)	∩d is [SP]
	□ o :	□(+S–P)	∩d is [S$\bar{P}$]
	◇ a :	◇(–S+P)	∪d is un[S$\bar{P}$]
	◇ e :	◇(–S–P)	∪d is un[SP]
	◇ i :	◇(+S+P)	∪d is [SP]
	◇ o :	◇(+S–P)	∪d is [S$\bar{P}$]

Note that the contradictory of □a is not □o but rather ◇o. Indeed □a, □e, ◇i and ◇o form a genuine square of opposition.

Figure I

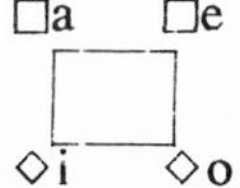

Given the following as axioms,[5]

Axiom 1. If $\Box x$ then x
Axiom 2. If x then $\Diamond x$

we can display the relations among our twelve assertoric and *de dicto* modal categoricals as this (where the arrow marks implication):

Figure II

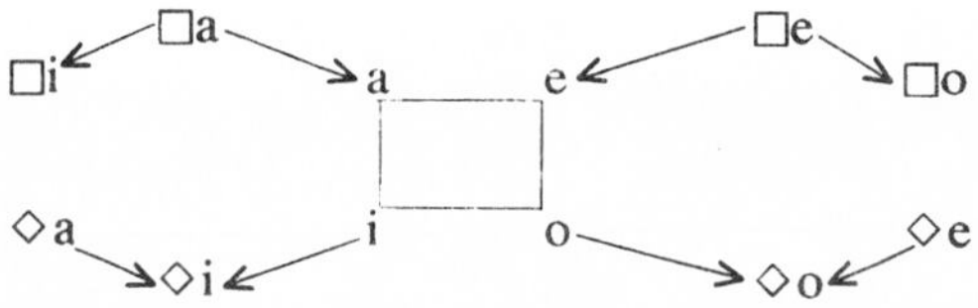

$\Box i \leftarrow \Box a \rightarrow a \qquad e \leftarrow \Box e \rightarrow \Box o$

$\Diamond a \rightarrow \Diamond i \leftarrow i \qquad o \rightarrow \Diamond o \leftarrow \Diamond e$

4. Unlike statements of *de dicto* modality, statements of *de re* modality are always made relative to a determinable domain (normally the actual world). Initially we can say that a statement of the form 'Some S is possibly P,' made relative to the actual world, simply claims that there are S's which are possibly P in the actual world. For example, 'Some horses might have wings,' made relative to the actual world, is first paraphrased as 'Some horses are possibly winged.' This statement implicitly claims that some possibly winged horses are constituents of the actual world. Our next step is to analyze the *de re* modality which persists here. When I state, relative to the actual world, that some horses are possibly winged I do not implicitly claim that the actual world has any winged horses. What I do implicitly claim is that there is some domain which has as constituents the (possibly winged) horses which are in the actual world, and these horses *are* winged in that other domain. Let us say that any domain which has as constituents the constituents of a given domain d is *accessible* from d (d is accessible to such a domain). If d is a domain, let $d*$ be a domain accessible from d. Now let $\cup d*$ be the union of domains accessible from d, and let $\cap d*$ be the intersection of domains accessible from d. We can say: $d \subseteq d*$. Accessibility is reflexive and transitive. Now, to state 'Some horses are possibly winged,' relative to

the actual world, a, is to claim implicitly that some domain accessible from a has a winged horse as a constituent.

Generally to make a *de re* problematic statement relative to a domain d is to claim implicitly that ∪d∗ has a specified (constitutive) property. To make a *de re* apodeictic statement relative to d is to claim implicitly that ∩d has a specified (constitutive) property. We display below the eight *de re* modalized categoricals along with their truth claims (where the relevant domain is d).

Table 2.

	Formula	Truth claim
a □:	–S+□P	∩d∗ is un[S$\bar{P}$]
e □:	–S+□–P	∩d∗ is un[SP]
i □:	+S+□P	∩d∗ is [SP]
o □:	+S+□–P	∩d∗ is [S$\bar{P}$]
a ◇:	–S+◇P	∪d∗ is un[S$\bar{P}$]
e ◇:	–S+◇–P	∪d∗ is un[SP]
i ◇:	+S+◇P	∪d∗ is [SP]
o ◇:	+S–◇–P	∪d∗ is [S$\bar{P}$]

Notice here that the contradictory of a□ is not o□ but o◇. The a□, e□, i◇ and o◇ formulae form a genuine square of opposition.

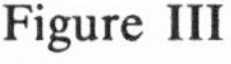
Figure III

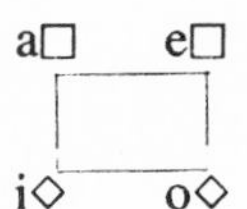

Given the following axioms;

Axiom 3. If x□ then x
Axiom 4. If x then x◇

we can display the relations among our twelve assertoric and *de re* modal categoricals as this.

Figure IV

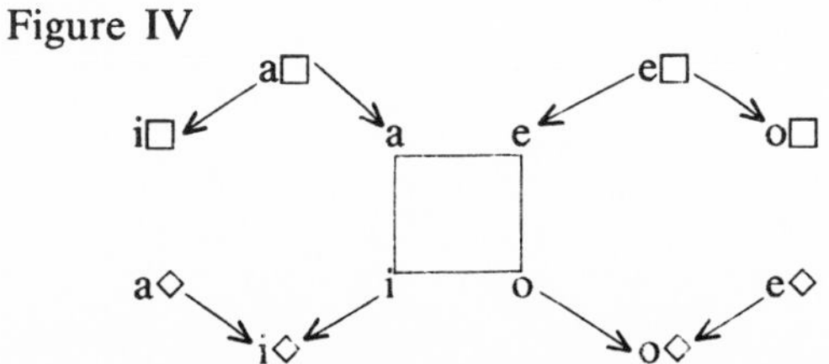

An examination of Tables 1 and 2 shows that, given that $\cup d{*}\subseteq\cup d$, and that $\cap d\subseteq\cap d{*}$ (since, in general, $A\subseteq A\cup B$ and $A\cap B\subseteq A$), all *de dicto* apodeictic statements entail their corresponding *de re* apodeictic statements, and all *de re* problematic statements entail their corresponding *de dicto* problematic statements. In other words

If □x then x□
If x◇ then ◇x

This means we could combine Figures II and IV to get the following diagram, which displays all the relations among our twenty assertoric and modalized categoricals.[6]

Figure V

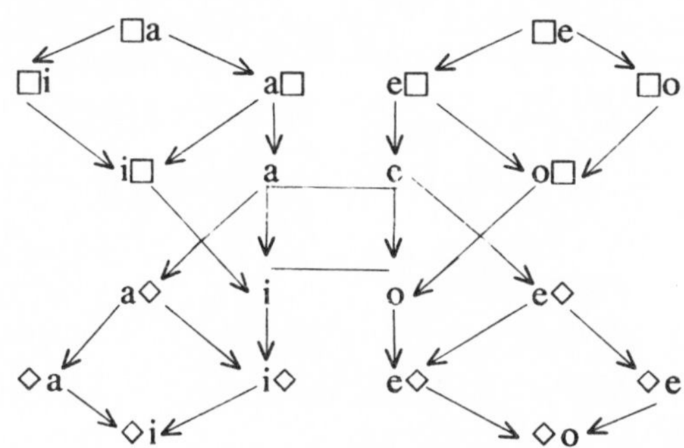

5. The number of assertoric syllogistic forms is quite large. But only about two dozen are actually valid. When we allow the modalization of terms and sentences these numbers increase dramatically. Restricting ourselves to syllogisms consisting only of sentences which are either assertoric or of *de dicto* modality, the number of syllogisms for each mood is twenty-seven. This is also the number when the modality is restricted to *de re*. So the total number of kinds of syllogisms for each mood is fifty-three. Of these several hundred syllogistic forms, fewer than half are actually valid.[7] Further syllogistic forms could be generated by permitting a mixture of *de re* and *de dicto* modality within the safe syllogism. Ockham claimed that by allowing such forms the number of valid moods is about one thousand (see Bochinski, (1948), pp. 229-230). In this section we will

look at the modalized Barbaras, specify some rules of proof, prove some of the valid Barbaras and, finally, lay down the conditions of validity for a modal syllogism.

The number of Barbaras with sentences which are either assertoric or of *de re* modality is twenty-seven. Twelve of these are valid. They are:

$$\begin{array}{llllll} 1.r & -M+\Box P & 2.r & -M+\Box P & 3.r & -M+\Box P \\ & -S+\Box M & & -S+\Box M & & -S+\Box M \\ \cline{2-2}\cline{4-4}\cline{6-6} & -S+\Box P & & -S+P & & -S+\Diamond P \end{array}$$

$$\begin{array}{llllll} 4.r & -M+\Box P & 5.r & -M+\Box P & 6.r & -M+\Box P \\ & -S+M & & -S+M & & -S+M \\ \cline{2-2}\cline{4-4}\cline{6-6} & -S+\Box P & & -S+P & & -S+\Diamond P \end{array}$$

$$\begin{array}{llllll} 7.r & -M+\Diamond P & 8.r & -M+P & 9.r & -M+P \\ & -S+\Box M & & -S+\Box M & & -S+\Box M \\ \cline{2-2}\cline{4-4}\cline{6-6} & -S+\Diamond P & & -S+P & & -S+\Diamond P \end{array}$$

$$\begin{array}{llllll} 10.r & -M+P & 11.r & -M+P & 12.r & -M+\Diamond P \\ & -S+M & & -S+M & & -S+M \\ \cline{2-2}\cline{4-4}\cline{6-6} & -S+P & & -S+\Diamond P & & -S+\Diamond P \end{array}$$

Note that 10.r is the standard Barbara, and 4.r is Barbara I(r) (cf. section above).

Proofs of the validity of each of these twelve moods requires, first, the usual rule of syllogistic inference--the *dictum de omni*. This rule, which demands that what is predicated of a distributed term is predicated of what that term is affirmed of, amounts, in Sommers' system, to the requirement that the algebraic sum of the premises must be equivalent to the conclusion. Secondly, axioms 3 and 4 (section four above) are required. Here are proofs for the first three syllogisms.

Proof of 1.r:

1.	$-M+\Box P$	premise
2.	$-S+\Box M$	premise
3.	$-S+M$	2, axiom 3
4.	$-S+\Box P$	1, 3, *dictum de omni*

Proof of 2.r:	1.	–M+□P	premise
	2.	–S+□M	premise
	3.	–M+P	1, axiom 3
	4.	–S+M	2, axiom 3
	5.	–S+P	3, 4, *dictum de omni*

Proof of 3.r:	1.	–M+□P	premise
	2.	–S+□M	premise
	3.	–S+M	2, axiom 3
	4.	–S+□P	1, 3, *dictum de omni*
	5.	–S+P	4, axiom 3
	6.	–S+◇P	5, axiom 4

It is easy to see that the logic of syllogisms containing one or more *de re* sentences is a simple extension (i.e., the addition of axioms 3 and 4) of the standard assertoric syllogistic. As Martha Kneale saw, these syllogisms simply have some special terms in them. The overall categorical forms remain. But the admission of *de dicto* modality into syllogisms requires a slightly more substantial extension of the assertoric syllogistic.

Of the twenty-seven Barbaras with sentences which are either assertoric or *de dicto* in modality only nine are valid. They are:

1.d	□(–M+P) □(–S+M) ——— □(–S+P)	2.d	□(–M+P) □(–S+M) ——— –S+P	3.d	□(–M+P) □(–S+M) ——— ◇(–S+P)
4.d	□(–M+P) –S+M ——— –S+P	5.d	□(–M+P) –S+M ——— ◇(–S+P)	6.d	–M+P □(–S+M) ——— –S+P
7.d	–M+P □(–S+M) ——— ◇(–S+P)	8.d	–M+P –S+M ——— –S+P	9.d	–M+P –S+M ——— ◇(–S+P)

Notice again that 8.d=10.r (the standard Barbara). Also note that while 4.r (Barbara 1(r)) was valid Barbara 1(d) is not among our nine valid moods above.

In addition to the *dictum de omni*, proofs for valid *de dicto* syllogisms require our axioms 1 and 2 (section three above), and the following rule:

Rule d: If □x and □y then □(x and y)

That this indeed should be a rule can be seen by looking at the truth claims of *de dicto* apodeictic statements. The claim of '□x' is '∩d is [x]'; that of '□y' is '∩d is [y].' Now in Sommers' system the following is a rule (in Sommers, (1982), it is Law 13(ii) on p. 185 and Theorem 12 on p. 401):

If X+A and X+B then X+<+A+B>

So, by this rule, if ∩d is [x] and ∩d is [y], then ∩d is <[x]+[y]>. And '∩d is <+[x]+[y]>' (i.e., 'the intersection of domains is both [x] and [y]') is the truth claim of '□(x and y).'

The following are proofs for the first three moods above.

Proof of 1.d:	1.	□(–M+P)	premise
	2.	□(–S+M)	premise
	3.	□(+[–M+P]+[–S+M])	1, 2, rule d
	4.	□(–S+P)	3, *dictum de omni*

Proof of 2.d	1.	□(–M+P)	premise
	2.	□(–S+M)	premise
	3.	□(+[–M+P]+[–S+M])	1, 2, rule d
	4.	□(–S+P)	3, *dictum de omni*
	5.	–S+P	4, axiom 1

Proof of 3.d:	1.	□(–M+P)	premise
	2.	□(–S+M)	premise
	3.	□(+[–M+P]+[–S+M])	1, 2, rule d
	4.	□(–S+P)	3, *dictum de omni*
	5.	–S+P	4, axiom 1
	6.	◇(–S+P)	5, axiom 2

Proofs similar to 1.r--3.r and 1d--3.d are available for modal versions of all valid moods. All that is needed to be added to the proof theory of the assertoric syllogistic are a small number of very simple principles (our four axioms and rule d). A system of greatly extended inference power is purchased for this small price. In fact, even the kinds of mixed valid syllogisms mentioned by Ockham are provable in our system. For example, we can easily prove:

□(–M+P)
–S+M
———
–S+◇P

Proof:

1.	$\Box$(–M+P)	premise
2.	–S+M	premise
3.	–M+P	1, axiom 1
4.	–S+P	2, 3, *dictum de omni*
5.	–S+$\Diamond$P	4, axiom 4

It has probably been noticed that apodeictic conclusions follow only from pairs of apodeictic premises. Indeed, we can think of our general forms of statements as arranged in order of "strength" (where one sentence is stronger than a second if and only if the first entails the second but the second does not entail the first). Our four axioms guarantee the following list of forms in descending order of strength:

$\Box$x
x$\Box$
x
x$\Diamond$
$\Diamond$x

Within each of these levels we have further levels determined by standard subalternation. Thus:

$\Box$a/e
$\Box$i/o
a/e$\Box$
i/o$\Box$
a/e
i/o
a/e$\Diamond$
i/o$\Diamond$
$\Diamond$a/e
$\Diamond$i/o

The relative orders of strength for statements is inversely proportional to the relative sizes of the domains which they denote--the stronger the statement the smaller the domain. For, given a domain d,

$$\cap d \subseteq \cap d* \subseteq d \subseteq \cup d* \subseteq \cup d$$

All this suggests that a necessary condition for the validity of *any* syllogism is that the conclusion cannot exceed any premise in strength. The scholastics called this the *peiorem* rule (*peiorem semper sequiter conclusio partem*).

Another necessary condition is suggested by the fact that no syllogism with both premises *de dicto* problematic or one premise and the conclusion *de dicto* problematic is valid. Let '$\Diamond$p' and '$\Diamond$q' be two of the sentences of a syllogism. Their truth claims are 'Ud is [p]' and 'Ud is [q]' respectively. Now 'Ud is [p]' is satisfied whenever at least one domain is [p]; and 'Ud is [q]' is satisfied whenever at least one domain is [q]. But there is nothing which guarantees that there is any common domain which is both [p] and [q]. In other words, though '◇p' and '◇q' are true just in case 'p' is true in some domain and 'q' is true in some domain, there is no guarantee that there is a domain in which they are both true. For example, both 'Possibly some horse is winged' and 'Possibly every winged object is a bird' are both true. Yet it need not follow that possibly some horse is a bird. This suggests that we demand of any valid syllogism that the respective domains of each of the sentences stand in the inclusion relation to one another.[8] Since this need not hold for the domains which ground the truth of any pair of *de dicto* problematic statements, we cannot permit more than one such statement within any syllogism. This, along with the *peiorem* rule, guarantees that if there is such a statement in a syllogism it must be the conclusion. If we add this new condition and the *peiorem* rule to the standard conditions for syllogistic validity we get the following five conditions, which are individually necessary and jointly sufficient for the validity of a syllogism.

1. The middle term must be distributed at least once.
2. Any term distributed in the conclusion must be distributed in the premises.
3. The number of particular conclusions must not be exceeded by the number of particular premises.
4. The conclusion must not exceed any premise in strength (*peiorem*).
5. The number of *de dicto* problematic sentences in a syllogism must not exceed one.[9]

We conclude this essay with a final cautionary note. Sommers has succeeded in building a version of syllogistic which matches (and may exceed) the standard predicate calculus in naturalness, expressive power and simplicity. We have attempted here to indicate one important direction in which his logic can be extended. That such an augmented syllogistic will have greatly expanded inference power is undeniable. Whether our own version of this extension is the only or best one is far from certain. We have offered only a glimpse of a new modal syllogistic--much more can and must be done.

Chapter IX

Endnotes

* This chapter first appeared in the *Notre Dame Journal of Formal Logic*, 29 (1988), and is reprinted here by permission of the Editors and the University of Notre Dame.

1. Aristotle's remarks allow for corresponding Darii versions as well.

2. Bochenski (1948), p. 58, took Barbara 1(d) to be valid. But he assimilated *de dicto* to *de re* modality.

3. There are restrictions on doing so however. See Sommers (1982), Chapter 14 and Chapter VI above.

4. For more on the status of singulars in the new syllogistic see Sommers (1969) and Chapter 6 of (1982), Englebretsen (1980a) and (1986a), Slater (1979a) and Wald (1979).

5. See *On Interpretation*, 22b11 and 23a21ff for Aristotle's statements of these.

6. Figure V should be seen as an improvement on McCall's Table 6 in (1963), p. 35.

7. McCall (1963), p. 46, claimed that 333 are valid when modality is restricted to *de dicto* or to *de re*.

8. This is actually part of a general condition on all discourse, requiring that there be some specifiable domain common to or included in all statements constituting that discourse. We cannot "mix domains." From facts about Pegasus and facts about Kentucky Derby winners we cannot draw conclusions about flying horses at the Derby.

9. Lukasiewicz took syllogisms with at least a *de dicto* problematic conclusion and major to be valid. See his (1957), p. 193.

Chapter X

EXISTING THINGS

Consider the two sentences, 'Something is red' and 'A red thing exists.' Are they synonymous? Given the standard first order predicate calculus, these are both formulated as '$(\exists x)Rx$.' So at least they have the same logical form according to orthodox theory. Indeed, for the modern orthodox logician, the logical role of the term 'exists' is fully absorbed by the particular quantifier (thus it is now called, appropriately, the "existential" quantifier). To be is to be the value of a bound variable. Bound variables (which appear as personal pronouns in natural language sentences) carry the burden of reference. To be able to refer to an object is to be able to countenance it as one of the values of a bound variable of a canonical sentence. To be able to so countenance it is to admit it to the universe of discourse, i.e., to grant its existence. Quine asked: What is there? and his answer was: Everything. This is what I will call the *first thesis* of standard mathematical logic.

First Thesis: Everything exists

The first thesis is accompanied by a *second thesis*. A sentence like 'Every singer is a performer' is reconstructed by the modern logician as having the form 'Everything is such that if it is a singer then it is a performer' ('$(\forall x)(Sx \supset Px)$'). Where the original English sentence seems to be about every singer, the new sentence seems to be about everything. Indeed, on the orthodox view, all general terms are construed as predicate terms, and their subjects are simply bound variables, pronouns, whose antecedents are 'something' or 'everything.' With the exception of those who introduce "sortal quantifiers," modern logicians are committed to an ontology of *things*--not dogs, persons, numbers, forms--things, bare particulars. "The variable is the legitimate latter-day embodiment of the incoherent old idea of a bare particular."[1] Thus:

Second Thesis: What exists are bare particulars

Our intent here is to cast doubt on both of these theses. Surprisingly, Sommers has continually paid lip service to the first thesis. He claims that existence is a characteristic not of things but of domains--a constitutive characteristic. It is, thus, more proper to say not that a ϕ-thing exists but that the appropriate domain is ϕ-ish. What Sommers is saying is that, given any thing, x, and any domain, D, x exists if and only if x is a constituent of D. What is at issue here is whether or not this notion does justice to our ordinary notion of existence (or at least our ordinary uses of 'exists'). Recall that every statement is made relative to some understood domain of discourse. The domain could be the actual world, the world of Greek mythology, the world of *Hamlet*, the set of all natural numbers, etc. For the statement to be true the domain must have as a constituent an individual satisfying the description determined by the statement. Now Sommers wants to hold that the existence of an individual and its status as a constituent of a given domain are analytically equivalent. To be is to be a member of a domain. But what, then, doesn't exist?

Suppose I wish to inform a child about Pegasus. I want to tell her that Pegasus is a winged horse. But I also want her to know that Pegasus does not exist. I might reasonably say, 'Pegasus is a winged horse which does not exist,' or 'Pegasus is a mythical, fictitious winged horse.' Given Sommers' theory of truth, what I say is true just in case the appropriate domain has as a constituent a winged horse named

'Pegasus' which does not exist. But, given that existence is to be analyzed as domain membership, it follows that for what I say to be true a winged horse named 'Pegasus' which does not exist must exist (i.e., be a constituent of the domain). Now I assume that the domain here is the world of Greek mythology. I might have even prefaced my statement with 'In Greek Mythology...' or 'In the stories and myths of ancient Greece...,' in order to make this clear to the child. It might be objected at this point that to say that my domain is a mythical world is to make redundant my statement that Pegasus does not exist. Either my domain is not a mythical world or I need not assert the nonexistence of any individual to which I refer. So perhaps my domain here is really something else. Surely it is not the actual world. But perhaps when I say 'x exists,' 'x does not exist,' 'x is real,' 'x is fictitious,' and the like, my domain must be construed as the omni-domain, the domain of all other domains. On this view, to say that 'x exists' is true is to say that x is a constituent of the omni-domain. Since an individual belongs to the omnidomain if and only if it belongs to some domain, 'x exists' will be true just in case x is a constituent of some domain. Indeed, on this view, everything exists.[2]

Now I don't believe that everything exists.[3] With the exception of a few landmarks, none of the individuals of Greek mythology exists. Hamlet does not exist. Santa does not exist. In a straightforward, common-sense way, what does exist are just constituents of the actual world. Tom Stoppard exists, but his Rosencrantz and Guildenstern do not. Stoppard is a constituent of the actual world--they are not. Of course, some things which appear in myths or stories do exist. But this is because they are not only constituents of their particular mythical or fictitious domains but of the actual world as well. Big Ben is a constituent of the actual world as well as Doyle's Sherlock Holmes stories.

There is a winged horse in Greek mythology. And there is a prince of Denmark named 'Hamlet' in Hamlet. If they do not exist then 'there is an x' or 'something is an x' cannot mean 'an x exists.' The distinction here, which Parsons ((1980), (1982)) has been at pains to reinforce, is one which, if ignored (as it is by most analytic philosophers today), leads to the Quine-Sommers' view that existence amounts to domain membership (though, admittedly, Quine and Sommers have different notions of domain). And this in turn leads to the first thesis. Consider one kind of dilemma faced by the defender of that thesis. Matt and Nat are two men, each sailing alone across the Atlantic. Matt is sailing for fun. He has no deadline, no particular destination or goal. Nat is searching for Atlantis. It would be natural and correct to describe Matt as not seeking anything. In response to the question 'What is Matt seeking?' we could say 'Matt is seeking nothing.' If asked the same question about Nat we must reply by making reference to Atlantis--'Nat is seeking Atlantis.' Nonetheless, counter-intuitively, the orthodox logician must hold that not only does Matt seek nothing, but Nat seeks nothing as well, since what he seeks is Atlantis, and Atlantis does not exist. Yet surely all that should logically follow from 'What Nat seeks is Atlantis' and 'Atlantis does not exist' is 'What Nat seeks does not exist.'

I think nothing is lost and much gained by abandoning the first thesis, and with it, therefore, the identification of existence with domain membership. One result of this abandonment is that statements of the forms 'Something is x' and 'An x exists' will sometimes have different truth-values. Statements of the second form (like those of such forms as '...does not exist,' '...is fictitious,' '...is real,' etc.) always have as their domain the actual world, and are true just in case the actual world has an x as a constituent. Statements like the former are made relative to a variety of domains (including the actual world), and their truth-values vary accordingly. 'There are winged horses' ('Something is a winged horse,' 'Some horse is winged') is true relative to the world of Greek mythology, but it is false relative to the actual world (and many others besides, e.g., the world of *Hamlet*).

It is important to notice that the rejection of the first thesis (and what that rejection entails) leaves Sommers' theory of truth intact. A statement is true just in case the domain relative to which the assertion is made has as a constituent an individual satisfying the description determined by the statement. We can even add that to say there is an x (something is an x) is to say that an x is a domain constituent. What we cannot add now, however, is that whatever is a domain constituent exists. Sommers' theory of truth has much to recommend it over alternative theories. Burdening it with superfluous theses about existence could well mean risking a sound ship for so much excess ballast.

What now of the second thesis? Ironically, among those who have argued against the notion of bare particulars (including Aristotle,[4] Frege[5] and Geach[6]) Quine seems to appear. For it is he who has earned wide acceptance for the view that there is "no entity without identity." No object is to be countenanced (as existing) unless an account of the conditions for identifying (and reidentifying) it is available, at least in principle. It is the lack of appropriate identity conditions which justifies Quinean qualms about intensional objects such as meanings, propositions and possibilia. Now Quine has said, "Identity is intimately bound up with the dividing of reference."[7] This is the kind of reference which general terms have. Thus, it follows that to countenance an entity (i.e., to admit it as an existent) is to be able to give an account of its conditions of identity in terms of some general term.

The general terms which are required for the accounting of identity conditions are *sortal* terms (what grammarians sometimes call "classificatory nouns" and Geach calls "substantivals"). To know *what* a thing is need not be to know *that* it is. What a thing is (what sort of thing it is) is only a necessary condition for knowing that it exists. Every thing is a thing of some sort.[8] No thing just is a thing without being any sort of thing. No thing is just a bare particular. The second thesis is false.

Terms like 'dog,' 'number,' 'idea,' and so forth, are sortal terms. We can pick out dogs, numbers and ideas--and we can count them.[9] On the other hand, there are terms which play the same grammatical roles as sortals but which are not themselves genuine sortals. They provide no means for identifying and reidentifying members of their extensions. The paradigm of these in English is 'thing' (others are 'object,' 'item,' 'individual,' 'entity'). We use a "pseudo-sortal" like 'thing' as a place-holder for a genuine sortal or for a disjunction of genuine sortals. When we say, for example, 'This thing is blocking my view' we and our audience both understand what this thing is. We both take it to be a clock, or curtain, or crown. So 'thing' will do, it simply holds the place for a genuine sortal. When we say 'Something is in the chimney' we don't know what sort of thing it is. But we do know a range of sorts to which it belongs. We know that it could be a bird or a bat, but not a number or a dump truck. We, in fact, probably take it to be some small animal, so that what we intend is something like 'Some bird, or bat, or squirrel, or mouse, or...is in the chimney.' Here the term 'thing' does duty for a disjunction of sortals.

In spite of the fact that a term like 'thing' is not a genuine sortal, it is a genuine term. It has all things of every sort in its extension and comprehends the feature shared by all things. When used in a sentence it signifies the property of being a thing and denotes all members of the domain of discourse. Now anything which has the property of being a man and the property of being short has the property of being a short man. If we let [P] stand for the property signified by 'P' when used in a sentence, and [PQ] for the property of being both [P] and [Q], then we can say that short men have the property [short man]. Consider now the expression 'short thing.' As used in a sentence it denotes things which have the property [short] and the property [thing]. Thus they have the property [short thing]. But, since 'thing' always signifies a property satisfied by all domain members, the property [short thing] just is the property [short]. For any term 'S,' [S thing]=[S].

Everything is a thing. And every thing is a domain member. But not everything exists. Every thing which is a member of the actual world exists. Thus, as we have seen, the two sentences

(a) Some things are ghosts

and

(b) Ghosts exist

are (contra logical orthodoxy) not synonymous. A sentence like (a) must always be used relative to some tacit, but determinable, domain, say D. Thus:

(a.1) (Some things are ghosts)$_D$

Its truth-condition is that D contain a member which has [ghost thing] (= [ghost]), i.e., that D contain a ghost. Suppose (a.1) is true. Then whether ghosts belonging to D exist or not depends upon which domain D is. For consider now (b). A sentence like (b), which uses a term like 'exist,' is always used relative to the actual world. What it says is that some things are ghosts, relative to the actual world, say W_A. Thus:

(b.1) (Some things are ghosts)$_{W_A}$

The truth-condition of (b.1) is that W_A contains a ghost. A sentence like 'There are ghosts' (or 'Some things are ghosts') is true in some domains (e.g., *Hamlet*) and false in others (presumably, the actual world). A sentence like 'Ghosts exist' just goes proxy for the sentence 'Some things are ghosts' used relative to the actual world. One consequence of our view, then, is that we can meaningfully use sentences of the form 'Some ϕ do not exist,' which is prohibited by the syntax of today's standard logic.

Chapter X

Endnotes

1. Quine (1980a), p. 25. Quine makes similar remarks in (1980), p. 165. For more on this see Englebretsen (1984h).

2. This view is one Sommers held long before he formulated his theory of truth. It is a view I have tried to attack and expunge from the rest of his program. See Englebretsen (1972c), (1974c) and (1975d) and Sarlet (1976).

3. I have argued for this in the places cited in note 2 above and my view has been strengthened by reading Rescher (1959), Routley (1966a) and Parsons (1980) and (1982).

4. See the interesting discussion in Durrant (1973).

5. See Frege (1959), especially pp. 59-66, where the notion is explored in terms of concepts.

6. Especially in Geach (1962), sections 31-34, and (1957), section 16.

7. Quine (1960), p. 115.

8. See Englebretsen (1978).

9. See Sommers (1970-71) and Englebretsen (1971a) and (1971c).

Chapter XI

A BRIEF NOTE ON PSYCHOLOGISM

In his 1983 Dawes Hicks lecture Jonathan Barnes made a mild plea in favor of the old idea that logic is, in some sense, expected to account for the competence (if not performance) of normal speakers to think logically, rationally.

> There is nothing disreputable in the old notion that logic studies the laws of thought--provided that the notion is not mistaken for the thesis that formal logic is concerned with the ways in which men actually think, and provided, too, that we do not suppose such study to exhaust the scope of logical research.[1]

In that same year Fred Sommers expressed optimism in the prospect of reforging the links between logic and psychology.

> There are signs that anti-psychologist bias of modern logic is on the wane. In part this change is due to the psychological realism of linguists like Chomsky who argue for a biological basis for universal grammar (including the grammar of the canonical fragment). There is, in any case, reason for confidence that a revitalized traditional formal logic that exploits the logical syntax of the natural languages will play the crucial role in determining the direction of that part of cognitive psychology that is concerned with universal grammar and with deduction as a mental process.[2]

By any standard, logic is something to be handled with care and caution. It is powerful and important. It is something to reckon with. Those who would turn to the resources of logic must do so with proper preparation or with expert guidance. Among those who have sought insight from logic have been linguists. They have asked logic to supply the 'logical forms' of natural language sentences. Now fundamental to any system of logic is a theory of logical syntax, a theory which accounts for how the logical forms of sentences are determined. If there are different logical systems it is likely there are different theories of logical syntax. And there are different logical systems. So those who seek the produce of logic ought to be aware of just who their supplier happens to be. You can get protein from both the truck farmer and the rancher. But in the one case you're likely to get spinach and in the other beef. Beef is fine--unless you want a salad. Unfortunately, the linguist today often seems aware of only one kind of supplier-one system of logic. To be fair, the fault lies with the logician. And the linguist is not the only one who goes to the logician and finds only one brand from which to choose. Cognitive psychologists also go to the logician's store.

Frege urged upon mathematicians and logicians an important distinction: the explanation for how a judgment is arrived at is not necessarily the grounds for the acceptance of the judgment as true. Often we hold beliefs on irrational grounds. Some believe Thatcher is a less dangerous leader than Bush on the grounds that she is a woman. Being able to explain how a belief comes to be held is not an ability to account for why it is (if it is) true. Frege took the first task to be a psychological one. He pleaded with mathematicians and logicians to keep these separate and to concentrate only on the latter task. He was reacting here to a mid-nineteenth-century attitude among many philosphers (most notably Mill) which took logic to be a part of psychology.[3] In Germany this attitude was typical of those who belonged to the naturalist and empiricist camps. In rejecting naturalism and empiricism in favor of a Kantian epistemology Frege also rejected the concomitant psychologism.[4] Ironically, contemporary analytic philosophers, who trace their logical insights back to Frege and join him in rejecting psychologism, tend to accept (under the influence of Hume, Russell and Carnap) a generally empiricist epistemology.

In today's schools of analysis the student is often taught that psychologism is the view that in accounting for the logical justification for any inference psychological considerations may play a role. Psychologism, in this guise, is of course false. But it's a straw man. No one ever seriously held this view. It certainly was not the view Frege rejected. For the view he rejected was, when properly understood (and extracted from its epistemological surroundings), true.

If cognitive psychology can be defined as the study of, *inter alia*, the grounds of human beliefs, and if some of our beliefs are held completely on formal grounds, and if logic is the science of such formal grounds (viz. an investigation into the formal constraints on inference), then, indeed, it is proper to think of logic as a part of psychology.[5] Suppose there is a formally valid inference of B from A and that I believe A because I believe B and also recognize the validity of the inference. A cognitive psychologist interested in the grounds of my belief could do no better than to seek an account offered by the formal logician. Granted not all (not even most, perhaps not even many) of our beliefs are held on grounds accountable for by formal logic, but some are. We can call these *rational* beliefs. The process of arriving at them might be called *reason*. Formal logic, then, is the study of the formal, necessary, universal constraints which operate on reason. Since our reason is manifested in what we say, logic can best be seen as a study of certain constraints on our language.

Of course the question naturally arises as to just what language one has in mind when saying that reason is manifested in what we say. For there is a genuine choice here: either reason is reflected in natural language, the language we all use; or it is reflected in an artificial language, a language constructed for a special purpose (such as mathematics). The choice between a naturalist or a constructionist approach to language was forced on logicians by Frege. For the most part logicians before him were content to construe logic as an attempt to formulate rules governing inferences made in natural languages. While some regimentation of natural language sentences was often called for, there was no doubt in the minds of traditional logicians that their logic was a logic of natural language. Frege's logical investigations began with his recognition that mathematical proof in his day tended to be informal and often permitted logical gaps. What he sought was a logical foundation for mathematics which would rectify this and bring formal rigor to mathematics. This was his program of logicism. What, in fact, Frege sought was a *Begriffschrift*, which in turn could serve in formulating the foundation for mathematics. He was convinced that in spite of shortcomings in proof procedures mathematics best represented human reasoning capacities. By contrast, he was convinced that natural language is imprecise, complex, cluttered, and in general a poor medium for reasoning. So, While preFregeans looked at natural language as the proper medium for reason, Fregean logicians held that reason could be properly carried out only in the medium of an artificially constructed, formal language. For preFregeans natural language is generally in good order; for Fregeans natural language is logically flawed. For preFregeans there is a logic of natural language; for Fregeans natural language either has no logic or it has a logic too complex to warrant investigation.

There is a science which deals with the formal structures of sentences in natural language--grammar. Grammar, as the science of natural language syntax, is one of the oldest sciences. Traditionally it was closely linked with formal logic. The grammarian sought criteria for specifying what was or was not a grammatically acceptable natural language sentence, and the logician sought criteria for specifying which inference relations existed among these sentences on the basis of their structure alone. In the sixteenth and seventeenth centuries 'philosophical' grammarians took the stronger position that there were criteria for grammaticality which were universal, common to all particular languages. The links between traditional formal logic, as simultaneously a *scientia rationalis* and a *scientia sermocinalis*, and philosophical grammar were obvious. The fruits of those links are best seen in the work of the Port Royal logicians.[6]

The eighteenth and early nineteenth centuries saw a gradual decline in both formal logic and philosophical grammar. Interest in the former was revived by the nineteenth-century algebraic logicians, such as Boole, DeMorgan and Schroeder; and it was firmly set in its present place by Frege, Russell and Whitehead. By contrast philosophical, or universal, grammar was overshadowed during most of this period by positivistic and empiricist fashions in linguistics (which forged links not with logic but with history, anthropology and sociology). The recovery of a philosophical grammar, which dominates today's schools of linguistic study, was due of course to Noam Chomsky, who claims ancestory not only among earlier linguists but among the sixteenth-and seventeenth-century rationalist philosophers and logicians, and philosophical grammarians.

There are two aspects of Chomsky's grammar which are of import for the remarks above on psychologism. First of all, the new grammar, like traditional grammar, is a grammar of natural language; and like the old philosophical grammar, it is a universal grammar. This means that modern-day Fregeans, such as Peter Geach, who suggest that Chomsky would do better to conform his account of sentence form to the dictates of Frege's logical syntax, are quite misguided. For as we have seen, a Fregean logic and a philosophical grammar of natural language are, *au fond*, incompatible. The second aspect of the new grammar is more important. Chomsky and his followers have claimed that the universal rules of natural language grammaticality are innate. Indeed, these rules are said to be species-specific, biologically determined predispositions which we all, normally, share from birth. Their reasons for holding this 'innateness hypothesis' are well known (the most important being the ability of the hypothesis to account for the rapid acquisition of language by children on the basis of a small and underdetermining series of model sentences). A consequence of the innateness program is that these grammarians conclude that the study of universal grammar is, in effect, an integral part of cognitive psychology (and in turn, perhaps even a part of neurophysiology). Thus, if the innateness hypothesis turns out to be viable, and if formal logic, *contra* Frege, is a logic of natural language, supplying the universal constraints on inferences among natural language sentences according to the forms of those sentences, then it follows that a logic of natural language must be an integral part of the study of some cognitive process (viz. reasoning).

When a logician says that 'Some mortal is human' follows formally from 'Some human is mortal' he or she is surely making a claim which, among other things, describes a correct process of reasoning. If so, then psychologism (but not the kind feared by contemporary Fregeans) has a useful and proper place in formal logic.

Chapter XI

Endnotes

1. Barnes (1983), p. 320, n.1.

2. Sommers (1983c), p. 44. See also Sommers (1978), Dauer (1981) and Peterson (1986).

3. A survey and discussion of the debate is found in Henle (1962).

4. For a recount of this see Sluga (1980), especially Chapter 2.

5. An excellent defense of this thesis (and much more of value) is found in Macnamara (1986).

6. For an interesting study of one of these links see Sommers (1983b).

BIBLIOGRAPHY

This bibliography consists of three parts. Part I is a list of works by Sommers. Part II is a list of works by the author relating in one way or another to syllogistic or to Sommers' philosophy in general. Part III contains all other works, concerning Sommers or otherwise, consulted or referred to in the text.

PART I

(1952), "The Passing of Privileged Uniqueness," *Journal of Philosophy*, 49.

(1959), "The Ordinary Language Tree," *Mind*, 68.

(1963), "Meaning Relations and the Analytic," *Journal of Philosophy*, 60.

(1963a), "Types and Ontology," *Philosophical Review*, 72; reprinted in Strawson (1967).

(1964), "Truth-functional Counterfactuals," *Analysis* (Sup.), 24.

(1964a), "A Program for Coherence," *Philosophical Review*. 73.

(1965), "Truth-value Gaps: A Reply to Mr. Odegard," *Analysis*, 25.

(1965a), "Predicability," *Philosophy in America*, M. Black, ed., Ithaca.

(1966), "Why Is There Something And Not Nothing?" *Analysis*, 26.

(1966a), "What We Can Say About God," *Judaism*, 15.

(1967), "On a Fregean Dogma," *Problems in the Philosophy of Mathematics*, I. Lakatos, ed., Amsterdam.

(1969), "Do We Need Identity?" *Journal of Philosophy*, 66.

(1969a), "On Concepts of Truth in Natural Languages," *The Review of Metaphysics*, 23.

(1970), "The Calculus of Terms," *Mind*, 79: reprinted in Englebretsen, (1987).

(1970-71), "Confirmation and the Natural Subject," *Philosophical Forum*, 2.

(1971), "Structural Ontology," *Philosophia*, 1.

(1973), "Existence and Predication," *Logic and Ontology*, M. Munitz, New York.

(1973a), "The Logical and the Extra-Logical," *Boston Studies in the Philosophy of Science*, 14.

(1975), "Distribution Matters," *Mind*, 84.

(1976), "Leibniz's Program for the Development of Logic," *Essays in Memory of Imre Lakatos*, R. S. Cohen, P. K. Feyerabend and M. W. Wartofsky, eds., Boston.

(1976a), "Frege or Leibniz?" *Studies on Frege*, III, M. Schirn, ed., Stuttgart.

(1976b), "Logical Syntax in Natural Language," *Issues in the Philosophy of Language*, A. MacKay and D. Merrill, eds., New Haven.

(1976c), "On Predication and Logical Syntax," *Language in Focus*, A. Kasher, ed., Dordrecht.

(1978), "The Grammar of Thought," *Journal of Social and Biological Structures*, 1.

(1978a), "Dualism in Descartes: The Logical Grounds," *Descartes*, M. Hooker, ed., Baltimore.

(1981), "Are There Atomic Propositions?" *The Foundations of Analytic Philosophy, Midwest Studies in Philosophy*, VI, P. A. French, T. E. Uehling, Jr. and H. K. Wettstein, eds., Minneapolis.

(1982), *The Logic of Natural Language*, Oxford.

(1983), "Reply to Geach," *Times Literary Supplement* (14 Jan.).

(1983a), "Reply to Geach," *Times Literary Supplement* (18 Feb.).

(1983b), "Linguistic Grammar and Logical Grammar," *How Many Questions? Essays in Honour of Sidney Morgenbesser*, L. Cauman, I. Levi, C. Parsons and R. Schwartz, eds., Indianapolis.

(1983c), "The Grammar of Thought: A Reply to Dauer," *Journal of Social and Biological Structures*, 6.

(1987), "Truth and Existence," in Englebretsen (1987).

Forthcoming, "Predication in the Logic of Terms," *Notre Dame Journal of Formal Logic*.

PART II

(1969), "Knowledge, Negation and Incompatibility," *Journal of Philosophy*, 66.

(1970), "Sommers' Tree Theory, Possibility and Existence," Ph.D. thesis, Nebraska.

(1971), "Elgood on Sommers' Rules of Sense," *Philosophical Quarterly*, 21.

(1971a), "Sommers' Theory and the Paradox of Confirmation," *Philosophy of Science*, 38.

(1971b), "On the Nature of Sommers' Rule," *Mind*, 80.

(1971c), "Nelson on Logical Notation," *Ratio*, 13.

(1972), "Vacuousity," *Mind*, 81.

(1972a), "On vanStraaten's Modification of Sommers' Rule," *Philosophical Studies*, 23.

(1972b), "True Sentences and True Propositions," *Mind*, 81.

(1972c), "Sommers on Empty Domains and Existence," *Notre Dame Journal of Formal Logic*, 13.

(1972d), "Persons and Predicates," *Philosophical Studies*, 23.

(1972e), "A Revised Category Mistake Argument," *Philosophical Studies*, 23.

(1973), "Persons, Predicates and Death," *Second Order*, 2.

(1973a), "Presupposition, Truth and Existence," *Philosophical Papers*, 2.

(1974), "Erwin on the Category Mistake Argument," *Second Order*, 3.

(1974a), "A Note on Contrariety," *Notre Dame Journal of Formal Logic*, 15.

(1974b), "Brody on Sommers," *Philosophical Studies*, 26.

(1974c), "Sommers on the Predicate 'Exists'," *Philosophical Studies*, 26.

(1975), *Speaking of Persons*, Halifax.

(1975a), "Sommers' Proof That Something Exists," *Notre Dame Journal of Formal Logic*, 16.

(1975b), "Trivalence and Absurdity," *Philosophical Papers*, 4.

(1975c), "Sommers' Theory and Natural Theology," *International Journal for the Philosophy of Religion*, 6.

(1975d), "Rescher on 'E!'," *Notre Dame Journal of Formal Logic*, 16.

(1976), "Sommers' Tree Theory and Possible Things," *Philosophical Studies* (Ire.), 24.

(1976a), "The Square of Opposition," *Notre Dame Journal of Formal Logic*, 17.

(1977), "Review: *Issues in the Philosophy of Language*," *Philosophical Studies* (Ire.), 25.

(1978), "Aristotle on the Subject of Predication," *Notre Dame Journal of Formal Logic*, 19; reprinted in Menne and Offenberger (1985).

(1979), "Notes on the New Syllogistic," *Logique et Analyse*, pp. 85-86.

(1980), "On Propositional Form," *Notre Dame Journal of Formal Logic*, 21; reprinted in Menne and Offenberger (1985).

(1980a), "Singular Terms and the Syllogistic," *The New Scholasticism*, 54.

(1980b), "Noncategorical Syllogisms in the Analytics," *Notre Dame Journal of Formal Logic*, 21.

(1980c), "Denotation and Reference," *Philosophical Studies* (Ire.), 27.

(1980d), "A Note on Predication," *Dialogue*, 19.

(1980e), "Bryant on Sommers," *Critica*, 7.

(1981), *Logical Negation*, Assen.

(1981a), *Three Logicians: Aristotle, Leibniz and Sommers and the Syllogistic*, Assen.

(1981b), "Predicates, Predicables, and Names," *Critica*, 8.

(1981c), "A Further Note on a Proof by Sommers," *Logique et Analyse*, 94.

(1981d), "On the Terms of Truth," *Philosophical Papers*, 10.

(1981e), "A Journey to Eden: Geach on Aristotle," *Grazer Philosophische Studien*, 14.

(1981f), "A Note on Identity, Reference, and Logical Form," *Critica*, 8.

(1982), "Do We Need Relative Identity?" *Notre Dame Journal of Formal Logic*, 23.

(1982a), "Aristotle and Quine on the Basic Combination," *The New Scholasticism*, 56.

(1982b), "Natural Syntax, Logical Syntax, and Translation," *Australian Logic Teachers Journal*, 6.

(1982c), "Leibniz on Logical Syntax," *Studia Leibnitiana*, 14.

(1982d), "Predication Old and New," *Critica*, 14.

(1982e), "Reference, Anaphora, and Singular Quantity," *Dialogos*, 41.

(1984), "Opposition," *Notre Dame Journal of Formal Logic*, 25.

(1984a), "Logical Form and Natural Syntax," *Indian Philosophical Quarterly*, 11.

(1984b), "Feldman and Sommers on Leibniz's Law," *Dialogos*, 43.

(1984c), "Logical Structure and Natural Syntax," *Journal of Social and Biological Structures*, 7.

(1984d), "Notes on Quine's Syntactical Insights," *Grazer Philosophische Studien*, 22.

(1984e), "Syntactical Complexity," *Philosophical Inquiry*, 6.

(1984f), "Quadratum Auctum," *Logique et Analyse*, 107.

(1984g), "Despre Logica Conjunctiiler Intrapropozitionale," *Revista de Filosofie*, 31.

(1984h), "Carrollian Things", *Jabberwocky*, 13.

(1985), "Defending Distribution," *Dialogos*, 45.

(1985a), "Geach on Logical Syntax," *The New Scholasticism*, 59.

(1985b), "On the Proper Treatment of Negative Names," *The Journal of Critical Analysis*, 8.

(1985c), "Zur philosophischen Interpretation der Logik: Ein weiter aristotelischer Dialog," in Menne and Offenberger (1985).

(1985d), "Negative Names," Philosophia, 15.

(1985e), "Logical Primitives," *Indian Philosophical Quarterly*, 12.

(1985f), "Semantic Considerations for Sommers Logic," *Philosophy Research Archives*, 11.

(1986), "Czezowski on Wild Quantity," *Notre Dame Journal of Formal Logic*, 27.

(1986a), "Singular/General," *Notre Dame Journal of Formal Logic*, 27.

(1986b), "On Some Alleged Semantic Correlations," *The New Scholasticism*, 60.

(1986c), "Quine on Aristotle on Identity," *Critica*, 17.

(1987), *The New Syllogistic* (ed.), New York.

(1987a), "Natural Syntax and Sommers' Theory of Logical Form," in Englebretsen (1987).

(1987b), "Logical Polarity," in Englebretsen (1987).

(1987c), "Subjects," *Studia Leinbnitiana*, 19.

(1988), "Preliminary Notes on a New Modal Syllogistic," *Notre Dame Journal of Formal Logic*.

(1988a), "La Theorie des categories de Sommers: une nouvelle introduction," *Dialogue*, 27.

(1989), "Formatives," *Notre Dame Journal of Formal Logic*, 30.

PART III

Altham, J. (1971), "Ambiguity and Predication," *Mind*, 80.

Angelelli, I. (1967), *Gottlob Frege and Traditional Philosophy*, Dordrecht.

_____ (1982), "Predication: New and Old," *Critica*, 14.

Angell, R. B. (1966), "Material Conditionals and Modern vs. Traditional Syllogistic and Immediate Inference," *Journal of Symbolic Logic*, 31.

Anonymous, (1982), "Critical Notice: *The Logic of Natural Language*," *Choice*.

Anton, J. P. (1957), *Aristotle's Theory of Contrariety*, London.

Aristotle (1941), *The Basic Works of Aristotle*, R. McKean, ed., New York.

_____ (1949), *Aristotle's Prior and Posterior Analytics*, W. D. Ross, ed., Oxford.

_____ (1963), *Categories and De Interpretatione*, J. L. Ackrill, ed., Oxford.

Ayer, A. J. (1952), "Negation," *Journal of Philosophy*, 49.

Bacon, J. (1966), "Natural Deduction Rules for Syllogistic," *Journal of Symbolic Logic*, 31.

_____ (1969), "Ontological Commitment and Free Logic," *Monist*, 53.

_____ (1987), "Sommers and Modern Logic," in Englebretsen (1987).

Baker, A. J. (1956), "Category Mistakes," *Australasian Journal of Philosophy*, 34.

_____ (1972), "Syllogistic With Complex Terms," *Notre Dame Journal of Formal Logic*, 13.

Balowitz, V. (1980), "The Rules of the Syllogism without Distribution," *International Logic Review*, 22.

_____ (1982), "A Method for Testing Validity," *The Newsletter on the Teaching of Philosophy*.

Banks, P. (1950), "On the Philosophical Interpretation of Logic: An Aristotelian Dialogue," *Dominican Studies*, 3; reprinted in Menne (1962).

Barnes, J. (1983), "Terms and Sentences: Theophrastus on Hypothetical Syllogisms," *Proceedings of the British Academy*, 69.

Bastable, P. K. (1975), *Logic: Depth Grammar of Rationality*, Dublin.

Bellert, I. (1984), "Interpretive Model for Linguistic Quantifiers," *Foundations of Logic and Linguistics*, London and New York.

Bennett, J. (1968), "Note on Sommers," *Journal of Symbolic Logic*, 32.

_____ (1971), "Note on Sommers," *Journal of Symbolic Logic*, 36.

Bigelow, J. C. (1978), "Critical Notice of *Issues in the Philosophy of Language*," *Linguistics and Philosophy*, 2.

Bird, O. (1964), *Syllogistic and its Extensions*, Englewood Cliffs, N.J.

Black, M. (1944), "Russell's Philosophy of Language," *The Philosophy of Bertrand Russell*, P. A. Schillp, ed., New York.

_____ (1959), "Language and Reality," *Proceedings of the American Philosophical Association*, 32.

Bochenski, I. M. (1948), "On the Categorical Syllogism," *Dominican Studies, 1*; reprinted in *Logico-Philosophical Studies*, A. Menne, ed., Dordrecht.

Brady, R. (1980), "Significance Range Theory," *Notre Dame Journal of Formal Logic*, 21.

Brentano, F. (1956), *Die Lehre vom richtig Urteil*, F. Meyer-Hillebrand, ed., Bern.

Brody, B. (1972), "Sommers on Predicability," *Philosophical Studies*, 23.

Bryant, J. (1979), "Negation, Denial and Possibility," *Critica*, 11.

Buchdahl, G. (1961), "The Problem of Negation," *Philosophy and Phenomenological Research*, 22.

Byerly, H. (1973), *A Primer of Logic*, New York.

Byrd, M. (1986), "Review: *The Logic of Natural Language*," International Studies in Philosophy, 18.

Castandeda, H. N. (1976), "Leibniz's Syllogistic-Propositional Calculus," *Notre Dame Journal of Formal Logic*, 17.

Caton, C. E. (1982), "Review: *Semantic and Conceptual Development* by F. Keil," *Language*, 58.

Chadwick, J. A. (1927), "On Propositions Belonging to Logic," *Mind*, 36.

_____ (1928), "Singular Propositions," *Mind*, 37.

Chandler, H. S. (1967), "Excluded Middle," *Journal of Philosophy*, 64.

_____ (1968), "Persons and Predicability," *Australasian Journal of Philosophy*, 46.

Chastain, C. (1975), "Reference and Context," *Language, Mind, and Knowledge*, K. Gunderson, ed., Minneapolis.

Chomsky, N. (1966), *Cartesian Linguistics*, New York and London.

_____ (1966a), *Topics in the Theory of Generative Grammar*, The Hague.

_____ (1968), *Language and Mind*, New York.

Church, A. (1965), "History of the Question of Existential Import," *Logic, Methodology and Philosophy of Science*, Y. Bar-Hillel, ed., Amsterdam.

Clark, M. (1980), *The Place of Syllogistic in Logical Theory*, Nottingham.

Cleeve, J. (1970), "Critical Notice of *Problems in the Philosophy of Mathematics*," *Ratio*, 12.

Cogan, R. (1976), "A Criticism of Sommers' Language Tree," *Notre Dame Journal of Formal Logic*, 17.

Corcoran, J. (1972), *Ancient Logic and Its Modern Interpretations*, Buffalo.

Cornman, J. W. (1967), "Language and Ontology," *The Linguistic Turn*, R. Rorty, ed., Chicago.

_____ (1968), "Types, Categories and Nonsense," *Studies in Logical Theory, American Philosophical Quarterly*, Monograph 2.

_____ (1970), "Categories, Grammar and Semantics," *Inquiry*, 13.

Couturat, L. (1903), *La Logique de Leibniz*, Paris.

Cowan, D. A. (1980), *Language and Negation*, San Mateo, California.

Czezowski, T. (1955), "On Certain Peculiarities of Singular Propositions," *Mind*, 64.

Dascal, M. (1976), "Language and Money," *Studia Leibnitiana*, 8.

Dauer, F. W. (1981), "Logical and Grammatical Forms: A Response to Sommers," *Journal of Social and Biological Structures*, 4.

DeMorgan, A. (1926), *Formal Logic*, London.

_____ (1966), *On the Syllogism and Other Logical Writings*, London.

DeSousa, R. B. (1966), "The Tree of English Bears Bitter Fruit," *Journal of Philosophy*, 63.

Diamond, C. (1981), "What Nonsense Might Be," *Philosophy*, 56.

Dipert, R. (1981), "Pierce's Propositional Logic," *Review of Metaphysics*, 34.

Drange, T. (1966), *Type Crossings*, The Hague.

_____ (1969), "Harrison and Odegard on Type Crossings," *Mind*, 78.

Dudman, V. (1976), "From Boole to Frege," *Studies on Frege*, vol. 1, I. M. Schirn, ed., Stuttgart.

Duerlinger, J. (1968), "Drawing Conclusions from Aristotelian Syllogisms," *Monist*, 52.

Dummett, M. (1967), "A Comment on 'On a Fregean Dogma'," *Problems in the Philosophy of Mathematics*, I. Lakatos, ed., Amsterdam.

Durrant, M. (1973), "Numerical Identity," *Mind*, 82.

Elgood, A. (1970), "Sommers' Rules of Sense," *Philosophical Quarterly*, 20.

_____ (1974), "Categories in Natural Language," *Philosophical Quarterly*, 24.

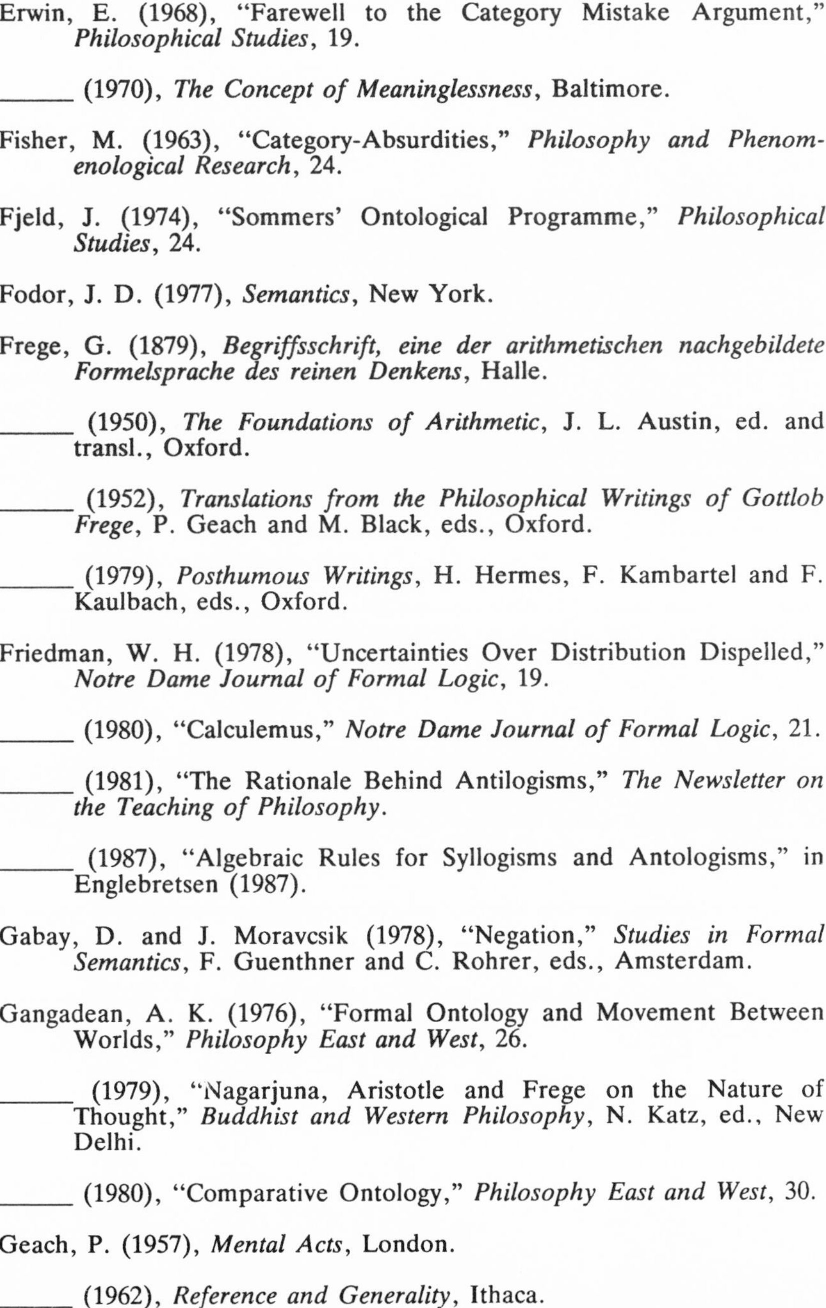

Erwin, E. (1968), "Farewell to the Category Mistake Argument," *Philosophical Studies*, 19.

_____ (1970), *The Concept of Meaninglessness*, Baltimore.

Fisher, M. (1963), "Category-Absurdities," *Philosophy and Phenomenological Research*, 24.

Fjeld, J. (1974), "Sommers' Ontological Programme," *Philosophical Studies*, 24.

Fodor, J. D. (1977), *Semantics*, New York.

Frege, G. (1879), *Begriffsschrift, eine der arithmetischen nachgebildete Formelsprache des reinen Denkens*, Halle.

_____ (1950), *The Foundations of Arithmetic*, J. L. Austin, ed. and transl., Oxford.

_____ (1952), *Translations from the Philosophical Writings of Gottlob Frege*, P. Geach and M. Black, eds., Oxford.

_____ (1979), *Posthumous Writings*, H. Hermes, F. Kambartel and F. Kaulbach, eds., Oxford.

Friedman, W. H. (1978), "Uncertainties Over Distribution Dispelled," *Notre Dame Journal of Formal Logic*, 19.

_____ (1980), "Calculemus," *Notre Dame Journal of Formal Logic*, 21.

_____ (1981), "The Rationale Behind Antilogisms," *The Newsletter on the Teaching of Philosophy*.

_____ (1987), "Algebraic Rules for Syllogisms and Antologisms," in Englebretsen (1987).

Gabay, D. and J. Moravcsik (1978), "Negation," *Studies in Formal Semantics*, F. Guenthner and C. Rohrer, eds., Amsterdam.

Gangadean, A. K. (1976), "Formal Ontology and Movement Between Worlds," *Philosophy East and West*, 26.

_____ (1979), "Nagarjuna, Aristotle and Frege on the Nature of Thought," *Buddhist and Western Philosophy*, N. Katz, ed., New Delhi.

_____ (1980), "Comparative Ontology," *Philosophy East and West*, 30.

Geach, P. (1957), *Mental Acts*, London.

_____ (1962), *Reference and Generality*, Ithaca.

_____ (1969), "Contradictories and Contraries," *Analysis*, 29; reprinted in Geach (1972).

_____ (1972), *Logic Matters*, Oxford.

_____ (1972a), "A Programme for Syntax," *Semantics of Natural Language*, D. Davidson and G. Harman, eds., Dordrecht.

_____ (1976), "Distribution and *Suppositio*," *Mind*, 85.

_____ (1982), "Review: *The Logic of Natural Language*," *Times Literary Supplement* (Nov. 26).

_____ (1983), "Reply to Sommers," *Times Literary Supplement* (Jan. 21).

Gensler, H. (1973), "A Simplified Decision Procedure for Categorical Syllogisms," *Notre Dame Journal of Formal Logic*, 14.

Gleitman, L. (1965), "Coordinating Conjunction in English," *Language*, 41.

Goddard, L. (1964), "Sense and Nonsense," *Mind*, 73.

_____ (1966), "Predicates, Relations and Categories," *Australasian Journal of Philosophy*, 44.

_____ (1970), "Nonsignificance," *Australasian Journal of Philosophy*, 48.

Goldstick, D. (1974), "Against Categories," *Philosophical Studies*, 26.

Greenberg, R. S. (1972), "Individuals and the Theory of Predication," *Journal of Philosophy*, 69.

Guerry, H. (1967), "Sommers' Ontological Proof," *Analysis*, 27.

Guttenplan, S. (1977), "Critical Notice: *Issues in the Philosophy of Language*," *Philosophical Books*, 18.

Haack, R. J. and S. Haack (1970), "Token-Sentences, Translation and Truth-Value," *Mind*, 79.

Haack, R. J. (1971), "No Need for Nonsense," *Australasian Journal of Philosophy*, 49.

Haack, S. (1968), "Equivocity: A Discussion of Sommers' Views," Analysis, 28.

Hallden, S. (1949), *The Logic of Nonsense*, Uppsala.

Hare, R. M. (1964), *The Language of Morals*, New York and Oxford.

Harrison, B. (1965), "Category Mistakes and Rules of Language," *Mind*, 74.

Henle, M. (1962), "On the Relation Between Logic and Thinking," *Psychological Review*, 69.

Hillman, D. J. (1963), "On Grammar and Category Mistakes," *Mind*, 72.

Hottois, G. (1978), "Critical Notice of Issues in the Philosophy of Language," *Revue Internationale de Philosophie*, 123.

Irwin, J. (1968), *On Categories and Types*, Ph.D. thesis, Syracuse.

Jong, W. R. de (1982), *The Semantics of John Stuart Mill*, Dordrecht.

Kalmar, L. (1967), "Not Fregean and not a Dogma," *Problems in the Philosophy of Mathematics*, I. Lakatos, ed., Amsterdam.

Kasher, A. (1972), "On Sommers' Concept of Natural Syntax," *Philosophical Studies* (Ire.), 20.

Kasher, A. and S. Lappin (1977), *Philosophical Linguistics*, Kronberg.

Keil, F. (1979), *Semantic and Conceptual Development*, Cambridge, Mass.

Kelly, C. J. (1985), "Two Neo-Kantian Arguments against Real Existential Predication," *Proceedings of the American Catholic Philosophical Association*, 59.

_____ (1986), "Aquinas' Third Way from the Standpoint of the Aristotelian Syllogism," *Monist*, 69.

Keynes, J. N. (1906), *Formal Logic*, London.

Khatchadourian, H. (1965), "Vagueness, Meaning, and Absurdity," *American Philosophical Quarterly*, 2.

King-Farlow, J. (1984), see Linsky and King-Farlow (1984).

Kneale, W. and M. (1962), *The Development of Logic*, Oxford.

Kopania, J. (1978), "Critical Notice: *Language in Focus*," *Studia Logica*, 37.

Lambert, K. (1968), "On the No-type Theory of Significance," *Australasian Journal of Philosophy*, 42.

Lappin, S. (1975), *Sortal Semantics*, Ph.D. thesis, Brandeis.

_____ (1977), see Kasher and Lappin (1977).

_____ (1978), "Sortal Incorrectness, Bivalence, and Classical Validity," *Manuscrito*, 2.

_____ (1981), *Sorts, Ontology, and Metaphor*, Berlin.

Largeault, J. (1975), "Critical Notice: *Logic and Ontology*," *Archives de Philosophie*, 38.

Lear, J. (1980), *Aristotle and Logical Theory*, Cambridge.

Leibniz, G. W. (1903), *Opuscules et Fragments inedits de Leibniz*, L. Couturat, ed., Paris.

_____ (1966), *Logical Papers*, G. H. R. Parkinson, ed., Oxford.

_____ (1968), *General Investigations Concerning the Analysis of Concepts and Truths*, N. H. O'Briant, ed., Athens, Ga.

Lejewski, C. (1967), "The Logical Form of Singular and General Statements," *Problems in the Philosophy of Mathematics*, I. Lakatos, ed., Amsterdam.

LePore, E. (1981), "Anaphoric Pronouns with Universal Quantifier Nominals as Antecedents," *Logique et Analyse*, 94.

Link, G. (1983), "Logical Semantics for Natural Language," *Erkenntnis*, 19.

_____ (1984), "Hydras: On the Logic of Relative Constructions with Multiple Heads," *Varieties of Formal Semantics*, F. Landman and F. Veltman, eds., Dordrecht.

Linsky, B. and J. King-Farlow (1984), "John Heintz's 'Subjects and Predicables'," *Philosophical Inquiry*, 6.

Lockwood, M. (1971), "Identity and Reference," *Identity and Individuation*, M. Munitz, ed., New York.

_____ (1975), "On Predicating Proper Names," *Philosophical Review*, 84.

_____ (1987), "Proofs and Pronouns: Extending the System," in Englebretsen (1987).

Long, D. C. (1983), "Review: *Descartes* by M. Hooker, ed.," *Nous*, 17.

Lukasiewicz, J. (1957), *Aristotle's Syllogistic from the Standpoint of Modern Formal Logic*, 2nd edition, Oxford.

MacKinnon, E. (1977), "Critical Notice: *Issues in the Philosophy of Language*," *Philosophy and Phenomenological Research*, 37.

Macnamara, J. (1986), *A Border Dispute: The Place of Logic in Psychology*, Cambridge, Mass.

Martin, J. J. (1975), "A Many-Valued Semantics for Category Mistakes," *Synthese*, 23.

Martin, Roger (1968), "Review: *Problems in the Philosophy of Mathematics*," *L'age de la Science*, 3.

Martin, R. L. (1967), "Toward a Solution to the Liar Paradox," *Philosophical Review*, 76.

_____ (1969), "Sommers on Denial and Negation," *Nous*, 3.

_____ (1969a), "Drange on Type Crossings," *Philosophy and Phenomenological Research*, 30.

Martin, R. M. (1958), *Truth and Denotation*, Chicago.

_____ (1979), *Pragmatics, Truth, and Language*, Dordrecht.

_____ (1983), *Mind, Modality, Meaning and Method*, New York.

_____ (1987), "On the Semantics of Sommers' 'Some S'," in Englebretsen (1987).

Martinich, A., et al. (1981), *Thomas Hobbes: Computatio Sive Logica*, New York.

Massey, G. "Tom, Dick, and Harry, and All the King's Men," *American Philosophical Quarterly*, 13.

Massie, D. (1967), "Sommers' Tree Theory: A Reply to DeSousa," *Journal of Philosophy*, 64.

Matthews, A. (1985), "Review: *The Logic of Natural Language*," *South African Journal of Philosophy*, 4.

Matthews, G. B. (1972), "Senses and Kinds," *Journal of Philosophy*, 69.

May, W. (1966), "Review: *Philosophical Logic*," *Philosophical Books*, 7.

Mayo, B. (1966), "Review: *Philosophical Logic*," *Philosophical Quarterly*, 16.

Maziarz, I. (1969), "Review: *Problems in the Philosophy of Mathematics*," *Philosophy of Science*, 36.

McCall, S. (1963), *Aristotle's Modal Syllogistic*, Amsterdam.

McCawley, J. "A Program for Logic," *Semantics of Natural Language*, D. Davidson and G. Harman, eds., Dordrecht.

McCulloch, G. (1984), "Frege, Sommers, Singular Reference," *Philosophical Quarterly*, 34; reprinted in *Frege: Tradition and Influence*, C. Wright, ed., Oxford.

McLaughlin, J. F. (1973), *Subject-Predicate and Logical Form*, Ph.D. thesis, Brandeis.

Medin, D. (1981), see Smith and Medin (1981).

Menne, A. (1950), "On the Philosophical Interpretation of Logic: An Aristotelian Dialogue," *Dominican Studies*, 3.

Menne, A. and N. Offenberger (1982), *Zur Modernen Deutung der Aristotelischen Logik*, vol. 1, Hildesheim and New York.

______ (1983), *Zur Modernen Deutung der AriStotelischen Logik*, vol. 2, Hildesheim and New York.

Merrill, K. R. (1982), "A Modest Defense of 'Bad Old Logic'," *The New Scholasticism*, 56.

Mill, J. S. (1843), *A System of Logic, Ratiocinative and Inductive*, London.

Moravcsik, J. (1978), see Gabay (1978).

Nelson, J. O. (1964), "On Sommers' Reinstatement of Russell's Ontological Program," *Philosophical Review*, 73.

______ (1965), "An Examination of Sommers Truth Functional Counterfactuals," *Theoria*, 31.

Noah, A. (1973), *Singular Terms and Predication*, Ph.D. thesis, Brandeis.

______ (1980), "Predicate Functors and the Limits of Decidability in Logic," *Notre Dame Journal of Formal Logic*, 21.

______ (1987), "The Two Term Theory of Predication," in Englebretsen (1987).

Novak, J. A. (1980), "Some Recent Work on the Assertoric Syllogistic," *Notre Dame Journal of Formal Logic,* 21.

Nuchelmans, G. (1966), "Review: *Philosophical Logic*," *Synthese*, 16.

Odegard, D. (1964), "On Closing the Truth-value Gap," *Analysis*, 25.

_____ (1966), "Absurdity and Types," *Mind*, 75.

Oesterle, J. A. (1963), *Logic: The Art of Defining and Reasoning*, Englewood Cliffs, N.J.

Offenberger, N. (1982), see Menne and Offenberger (1982).

_____ (1983), see Menne and Offenberger (1983).

Paduceva, E. V. (1970), "Anaphoric Relations and Their Representation in the Deep Structure of a Text," *Progress in Linguistics*, M. Bierwisch and K. Heidolph, eds., The Hague.

Pap, A. (1948), "Logical Nonsense," *Philosophy and Phenomenological* Research, 9.

_____ (1960), "Types and Meaninglessness," *Mind*, 69.

Parsons, T. (1977), "Type Theory and Ordinary Language," *Linguistics, Philosophy, and Montague Grammar*, S. Davis and M. Mithun, eds., Austin.

_____ (1980), *Non-existent Objects*, New Haven.

_____ (1982), "Are There Nonexistent Objects," *American Philosophical Quarterly*, 19.

Passell, D. (1969), "On Sommers' Logic of Sense and Nonsense," *Mind*, 78.

Patzig, G. (1968), *Aristotle's Theory of the Syllogism*, J. Barnes, transl., Dordrecht.

Peirce, C. S. (1933), *Collected Papers*, London.

Peterson, P. L. (1984), "Contrariety and the Cube of Opposition," (Abstract), *Journal of Symbolic Logic*, 49.

_____ (1986), "Real Logic in Philosophy," *Monist*, 69.

Post, J. F. (1970), "The Possible Liar," *Nous*, 4.

_____ (1973), "Shades of the Liar," *Journal of Philosophical Logic*, 2.

Potter, K. (1964), "Negation, Names and Nothing," *Philosophical Studies*, 15.

Presley, C. F. (1962), "Arguments About Meaninglessness," *British Journal for the Philosophy of Science.*

Prior, A. N. (1949), "Categoricals and Hypotheticals in George Boole and his Successors," *Australasian Journal of Philosophy*, 27.

_____ (1953), "The Logic of Negative Terms in Boethius," *Franciscan Studies*, 13.

_____ (1953a), "Negative Quantifiers," *Australasian Journal of Philosophy*, 31.

_____ (1956), "Logicians at Play; or Syll, Simp and Hilbert," *Australasian Journal of Philosophy*, 34.

_____ (1959), "Formalised Syllogistic," *Synthese*, 11.

_____ (1976), *The Doctrine of Propositions and Terms*, London.

_____ (1976a), *Papers in Logic and Ethnics*, London.

Purtill, R. (1971), *Logic for Philosophers*, New York.

_____ (1972), *Logical Thinking*, New York.

_____ (1979), *Logic: Argument, Refutation, and Proof*, New York.

_____ (1987), "Some Practical and Theoretical Features of Sommers' Cancellation Method," in Englebretsen (1987).

Quine, W. V. O. (1960), *Word and Object*, Cambridge, Mass.

_____ (1966), "Variables Explained Away," *Selected Logic Papers*, New York.

_____ (1967), "Three Remarks," *Problems in the Philosophy of Mathematics*, I. Lakatos, ed., Amsterdam.

_____ (1970), *Philosophy of Logic*, Englewood Cliffs, N.J.

_____ (1972), *Methods of Logic*, 3rd edn., New York.

_____ (1976), "Algebraic Logic and Predicate Functors," *The Ways of Paradox and Other Essays*, New York.

_____ (1976a), "The Variable," *The Ways of Paradox and Other Essays*, New York.

_____ (1980), "The Variable and its Place in Reference," *Philosophical Subjects*, Z. van Straaten, ed., Oxford.

_____ (1980a), "Grammar, Truth, and Logic," *Philosophy and Grammar*, S. Kanger and S. Ohman, eds., Dordrecht.

_____ (1981), *Theories and Things*, Cambridge, Mass.

_____ (1981a), "Predicate Functors Revisited," *Journal of Symbolic Logic*, 46.

Reardon, M. (1982), "On Teaching Students Logic," *Philosophy*, 57.

Reinhardt, L. R. (1965-66), "Dualism and Categories," *Proceeding of the Aristotelian Society*, 66.

Rescher, N. (1959), "On the Logic of Existence and Denotation," *Philosophical Review*, 68.

Richmond, S. A. (1971), "Sommers on Predicability," *Journal of Philosophy*, 68.

_____ (1975), "A Possible Empirical Violation of Sommers' Rule for Enforcing Ambiguity," *Philosophical Studies*, 28.

Rorty, R. (1976), *The Linguistic Turn*, Chicago.

Rosenbaum, S. E. (1972), *Properties and Categories*, Ph.D. thesis, Brown.

Rosenthal , D. M. (1976), "Possibility, Existence and an Ontological Argument," *Philosophical Studies*, 30.

Routley, R. (1966), "On a Significance Theory," *Australasian Journal of Philosophy*, 44.

_____ (1966a), "Some Things Do Not Exist," *Notre Dame Journal of Formal Logic*, 7.

_____ (1979), *Exploring Melnong's Jungle and Beyond*, Sydney.

Russell, B. (1940), *An Inquiry Into Meaning and Truth*, London.

_____ (1956), *Logic and Knowledge, Essays 1901-1950*, R. C. Marsh, ed., London.

_____ (1960), *Our Knowledge of the External World*, New York.

Ryle, G. (1949), *The Concept of Mind*, London.

_____ (1955), "Categories," *Logic and Language*, 2nd series, A. Flew, ed., Oxford.

Sanford, D. (1968), "Contraries and Subcontraries," *Nous*, 2.

Sarlet, H. (1976), "La Formalisation de 'Exist'," *Logique et Analyse*, 74-76.

Sayward, C. (1976), "A Defense of Sommers," *Philosophical Studies*, 29.

_____ (1978), "Strawson on Categories," *The Journal of Critical Analysis*, 7.

_____ (1981), "The Tree Theory and Isomorphism," *Analysis*, 41.

_____ (1987), "Some Problems With TFL," in Englebretsen (1987).

Sayward, C. and S. Voss (1972), "Absurdity and Spanning," *Philosophia*, 2.

_____ (1980), "The Structure of Type Theory," *Journal of Philosophy*, 77.

Sharpe, R. (1967), "Category Mistakes and Classification," *Inquiry*, 10.

Shiman, P. L. (1984), "Toward a Canonical Term Functor Logic," (Abstract), *Journal of Symbolic Logic*, 49.

Slater, B. H. (1972), "The Foundations of Logic," *Mind*, 81.

_____ (1973), "Logic," *Philosophical Quarterly*, 23.

_____ (1974), "Logic and Grammar," *Philosophical Quarterly*, 23.

_____ (1978), "A Fragment of a New Propositional Logic," *International Logic Review*, 9.

_____ (1979), "Wittgenstein's Later Logic," *Philosophy*, 54.

_____ (1979a), "Singular Subjects," *Dialogue*, 18.

_____ (1979b), "Aristotle's Propositional Logic," *Philosophical Studies*, 36.

_____ (1979c), "Internal and External Negation," *Mind*, 88.

_____ (1980), "Expressions of Ignorance," *Australasian Journal of Philosophy*, 38.

_____ (1981), "Direct Tableaux Proofs," *Analysis*, 41.

_____ (1982), "Incomplete Assertions," *Studia Logica*, 41.

_____ (1987), "Back to Leibniz or on from Frege?" in Englebretsen (1987).

Sluga, H. (1980), *Gottlob Frege*, London.

Smiley, T. (1961), "Syllogisms and Quantification," *Journal of Symbolic Logic,* 26.

_____ (1967), "Mr. Strawson on Traditional Logic," *Mind*, 76.

_____ (1973), "What is a Syllogism?" *Journal of Philosophical Logic*, 2.

Smith, E. and D. Medin (1981), *Categories and Concepts*, Cambridge, Mass.

Solomon, A. (1968), "Review: Problems in the Philosophy of Mathematics," *British Journal for the Philosophy of Science*, 19.

Sosa, E. (1973), "What is a Logical Constant?" *Boston Studies in the Philosophy of Science*, 14.

Stahl, G. (1964), "Linguistic Structures Isomorphic to Object Structures," *Philosophy and Phenomenological Research*, 24.

_____ (1969), "Formal Logic and Natural Languages," *Foundations of Language*, 5.

_____ (1983), "Revue: The Logic of Natural Language," *Revue Philosophique de la France et de l'Etranger*, 173.

Standley, G. (1974), "Review: 'On a Fregean Dogma'," *Journal of Symbolic Logic*, 39.

Stern, K. (1969), "Linguistic Restrictionism and the Idea of Potential Meaning," *Monist*, 53.

_____ (1977), "Critical Notice of Issues in the Philosophy of Language," *Philosophical Quarterly*, 27.

Stevenson, L. (1977), "Review: *Issues In the Philosophy of Language*," *Philosophical Quarterly*, 27.

Stoothoff, R. (1982), "Critical Notice: *Studies on Frege*, III," *Journal of Symbolic Logic*, 47.

Strawson, P. F. (1950), "On Referring," *Mind*, 59.

______ (1952), *Introduction to Logical Theory*, London.

______ (1959), *Individuals*, London.

______ (1967), *Philosophical Logic* (ed.), Oxford.

______ (1970), "The Asymmetry of Subjects and Predicates," *Language, Belief and Metaphysics*, M. Munitz and H. Kiefer, eds., Albany.

______ (1971), *Logico-Linguistic Papers*, London.

______ (1974), *Subjects and Predicates in Logic and Grammar*, London.

______ (1982), "Review: *The Logic of Natural Language*," *Journal of Philosophy*, 79; reprinted in Englebretsen (1987).

Suzman, J. (1972), "The Ordinary Language Lattice," *Mind*, 81.

Swanson, J. W. (1967), "Denial in First Order Logic," *Analysis*, 27.

Swiggart, P. (1972), "The Limits of Statement Denial," *Mind*, 81.

______ (1987), "DeMorgan and Sommers," in Englebretsen (1987).

Tarski, A. (1943-44), "The Semantic Conception of Truth and the Foundations of Semantics," *Philosophy and Phenomenological Research*, 4.

______ (1956), "The Concept of Truth in Formalized Languages," *Logic, Semantics, Metamathematics*, J. H. Woodger, ed., London.

Tennant, N. (1977), "Critical Notice: *Language in Focus*," *Philosophical Quarterly*, 27.

Thom, P. (1981), *The Syllogism*, Munich.

Thompson, M. (1957), "On Category Differences," *Philosophical Review*, 66.

VanBenthem, J. (1983), "Critical Notice: *The Logic of Natural Language*," *Philosophical Books*, 24.

______ (1984), "A Linguistic Turn: New Directions in Logic," *Proceedings of the Seventh International Congress on Logic, Methodology and Philosophy of Science*, Amsterdam.

Van Straaten, R. (1968), "Sommers' Rule and Equivocity," *Analysis*, 29.

_____ (1971), "A Modification of Sommers' Rule," *Philosophical Studies*, 22.

Veatch, H. (1950), "Aristotelian and Mathematical Logic," *The Thomist*, 13.

_____ (1952), "The Significance of the Current Criticisms of the Syllogistic," *The Thomist*, 15.

_____ (1968), "A Modest Word in Defense of Aristotle's Logic," *Monist*, 52.

Vendler, Z. (1967), *Linguistics in Philosophy*, Ithaca.

Venn, J. (1881), *Symbolic Logic*, London.

Verburg, D. (1969), "Hobbes' Calculus of Words," *Review of Applied Linguistics*, 5.

Verma, R. R. (1978), "Denial, Contradiction, and Truth-Value Gaps," *Philosophia*, 8.

Voss, S. (1972), see Sayward and Voss (1972).

_____ (1980), see Sayward and Voss (1980).

Wald, J. (1979), "Geach on Atomicity and Singular Propositions," *Notre Dame Journal of Formal Logic*, 20.

Watson, D. (1973), "On Types and Ontology," *Second Order*, 2.

Wiggins, D. (1967), *Identity and Spatio-Temporal Continuity*, Oxford.

Williamson, C. (1971), "The Traditional Logic as a Logic of Distribution-values," *Logique et Analyse*, 56.

Wittgenstein, L. (1961), *Tractatus Logico-Philosophicus*, D. Pears and B. McGuines, transl., London.

Woods, J. (1969), "Predicate Ranges," *Philosophy and Phenomenological Research*, 30.

Zemach, E. (1981), "Names and Particulars," *Philosophia*, 10.

INDEX OF NAMES

N

O

P

Q

R

S

T

U

V

W

Z

INDEX OF TERMS

D

E

F

G

I

L

M

N

O

P

Q

R

S

T

U

V

W

PROBLEMS IN CONTEMPORARY PHILOSOPHY

1. Petra von Morstein, **On Understanding Works of Art: An Essay in Philosophical Aesthetics**

2. David and Randall Basinger, **Philosophy and Miracle: The Contemporary Debate**

3. Franscico Peccorini Letona, **Selfhood as Thinking Thought in the Work of Gabriel Marcel**

4. Corbin Fowler, **The Logic of U.S. Nuclear Weapons Policy: A Philosophical Analysis**

5. Marcus P. Ford (ed.), **A Process Theory of Medicine: Interdisciplinary Essays**

6. Lars Aagaard-Mogensen (ed.), **The Idea of the Museum: Philosophical, Artistic, and Political Questions**

7. Kerry S. Walters, **The Sane Society in Modern Utopianism: A Study in Ideology**

8. Steven W. Laycock, **Foundations for a Phenomenological Theology**

9. John R. Jacobson and Robert Lloyd Mitchell (eds.), **Existence of God: Essays from the Basic Issues Forum**

10. Richard J. Connell, **The Empirical Intelligence - The Human Empirical Mode: Philosophy As Originating in Experience**

11. Sander H. Lee (ed.), **Inquiries into Values: The Inaugural Session of the International Society for Value Inquiry**

12. Tobias Chapman, **In Defense of Mystical Ideas: Support for Mystic Beliefs from a Purely Theoretical Viewpoint**

13. Donald Stewart, **Entities and Individuation: Studies in Ontology and Language**

14. Peter Preuss, **Reincarnation**

15. Tibor R. Machan, **The Moral Case for the Free Market Economy: A Philosophical Argument**

16. George Frederick Schueler, **The Idea of a Reason for Acting: A Philosophical Argument**

17. William and Harriet Lovitt, **Modern Technology in the Heideggerian Perspective**

18. William Cooney (ed.), **Reflections on Gabriel Marcel: A Collection of Essays**

19. Mari Sorri and Jerry Gill, **A Post-Modern Epistemology: Language, Truth, and Body**

20. Adolf Portmann, **Essays in Philosophic Zoology**: ***The Living Form and the Seeing Eye***, Richard B. Carter (trans.)

21. George Englebretson, **Essays on the Philosophy of Fred Sommers: In Logical Terms**

22. Kevin Doran, **What is a Person: The Concept and the Implications for Ethics**

23. Ronald Roblin (ed.), **The Aesthetics of the Critical Theorists: Studies on Benjamin, Adorno, Marcuse, and Habermas**

24. William Lane Craig (ed.), **The Logic of Rational Theism: Exploratory Essays**

25. Barrie A. Wilson, **Hermeneutical Studies: Dilthey, Sophocles, and Job**